By M H BASLOW

The Evolution of Mister Newman II

A Story of Mankind Faced with the Sudden Appearance of a Reasoning and Intellectually Gifted Animal Ancestor

The Evolution of Mister Newman
II

Copyright © 2013 by M H BASLOW
Addendum: Chapter 14, 2014

All rights reserved. No part of this book may be reproduced or transmitted in any form or by any means without written permission of the author.

Library of Congress Control Number: 2012954303

ISBN: 978-1500998042

WOODLAND ROAD PRESS, NEW YORK

Enquiries: woodlandroadpress@gmail.com

Printed in the United States of America

To My Family

Tracy, Lisa and Ronald and in memory of Trudy

Table of Contents

Foreword..i

Prologue..1

Chapter 1 Attack on the Citadel..........................3

Chapter 2 The Cave......................................19

Chapter 3 Resurrection..................................23

Chapter 4 Genesis.......................................29

Chapter 5 The Early Years...............................51

Chapter 6 An Extraordinary Person.......................59

Chapter 7 The Encounter.................................71

Chapter 8 The Wedding...................................79

Chapter 9 Family..85

Chapter 10 Diaspora......................................93

Chapter 11 Revelation...................................105

Chapter 12 The Gathering................................117

Chapter 13 Trial..123

Chapter 14 Captivity....................................167

Epilogue..215

Foreword

In 1925, a teacher named John T. Scopes was tried for falsely teaching, based on Darwin's theory of evolution that men had evolved from and were therefore related to the apes. Two prominent lawyers of that time participated in a "show" trial pitting science against religious dogma. Mr. William Jennings Bryan prosecuted the case against Mr. Scopes, and Mr. Clarence S. Darrow was the lawyer for the defense. The verdict went against Mr. Scopes and this case, widely known as the Scopes "Monkey Trial", stands as an example of the occasional and irrational intransigence of humanity in the face of overwhelming scientific evidence.

The fictional story entitled "The Evolution of Mister Newman" involves a series of events set in motion by the Scopes trial that led to the opening of a similar trial many decades later - this time however, with the presence of a key witness for the defense missing at the earlier trial. Who this mysterious witness was, and how he came into being is the essence of the story, a story that is presented to the reader as told by his eldest son Peter.

Whereas this story contains many examples of human kindness and understanding, it also involves elements of human frailty, including criminal assault, animal abuse, murder and revenge.

Foreword to: "The Evolution of Mister Newman II"

In this expanded version of the book, a new chapter, Chapter 14, entitled "Captivity" has been added which covers a period of time after the trial and during which Mister Newman was held in captivity and studied extensively by the medical community. In the original version, Mister Newman was described as an intellectually gifted individual but his gifted nature and his philosophy were not fully developed in order to advance the story line.

In chapter 14, we get a chance to learn about his unique philosophy. His philosophy is revealed to us during discussions with his friend and champion, Dr. Paula Chapman. The topics covered include the creation of the universe; the nature of our Creator; the existence of a soul; free will; the genetic code; the origin of life on earth; the nature of memories and of the mind, and the future of our creation. Paula and Mister Newman propose a new religion, *"Naturalism"* and a new *Bible* that is based on natural science rather than metaphysics. They also note that there are already many practicing *Naturalists* within the various religions of mankind as well as among atheists.

Prologue

My name is Peter and I am the eldest son of Mister Mangor Newman.

This story that I have entitled "The Evolution of Mister Newman" is written as a testament to my father's life and teachings while he was among us, and although written long after my father was gone I believe that it is an accurate account of the events as they unfolded. The story is based on first-hand discussions with my father and on interviews with others who were involved in this matter at the time of his trial. In addition to those interviews, it is based on the contents of several existing records including six journals written by my grandfather and on the extensive court transcripts in this case. Lastly, it is also based on my personal recollections as a child, and on stories told to me by my grandfather about my father's time among us before I was born. The story is based on available factual information as well as on my reconstruction of some of the events as they probably happened. While it is hard to decide just where to start, I think it best to start with the day that my father's existence and special nature became known to the world at large.

Although there are many readers who may choose not to believe this narrative for a variety of reasons, for me and

Prologue

my brothers who have lived through this time, there is no question about the truth of this story as presented. The following list of the main characters in the story may be useful to the reader as a reference in following elements of the story as they unfold.

Mister M. Newman........................ *my father*
Mary White…………………………..*my mother*
Dr. Joseph Newman, DVM…………..*my grandfather*
A stranger born in Uganda, Africa……*my grandmother*
Peter M. Newman…………………….*myself*
Matthew M. Newman…………………..*my younger brother*
Andrew M. Newman…………………..*my youngest brother*
Roger Malek…………………………..*the hunter*
Alan Garfield…………………………..*the sheriff*
Dr. Paula Chapman, PhD……………..*the professor*
Mr. Henry Dawson…………………….*the judge*
Mr. John Bryan………………………..*lawyer for the State*
Mr. Raymond Darrow………….……...*lawyer for Mangor*

Chapter 1
Attack on the Citadel

The winter's snows had been especially heavy this year, and the cold and the dampness had made his joints stiffer and even more painful than usual. However, on awakening this morning he sensed a change. Outside, the sun was shining, the sky was bright and blue, and in the quiet of his mountain retreat he could hear the melted snow falling from the rocks above.

"Today will be a fine day," thought the old one as he prepared the morning fire.

In each and every year, there are some special days that seem to fit into no season, but instead have a magic all their own. Although they cannot be predicted with any certainty, when they arrive there can be no mistaking them. This was one of those days; days that fill one with a sense of the changing of the seasons and of the promise of God's works about to unfold. In the high mountain meadow where the old one lived, it happened each year about the beginning of the month of May when the sun had reached high enough into the sky to warm the cliffs and forests that bordered the meadow. He felt warm and happy and as he prepared his morning's tea, visions of deep green woods and the smells of summer yet to come crowded into his senses. On this

Chapter 1 Attack on the Citadel

particular day even his aged body threw off its discomforts and was in harmony with both heaven and earth.

After breakfast he cleaned up, put away the dishes and then went outside to greet the day. The old one lived an isolated and lonely life high in the Sierras. He saw people rarely and when he did he carefully avoided any contact with them. His home was at the base of some cliffs that rose steeply from the northernmost portion of the meadow below and today, feeling as well as he did, the old one climbed the ridge in the bright warm sunlight and stood with outstretched arms looking out at his universe. This was indeed his universe. There were few others, man or beast, who could survive in it. He alone had mastered it here high among the loftiest Sierra peaks. Let nature give him her best or her worst – he had conquered the wind, the cold, the snow and hail, and the winters' darkness. What a sense of achievement he had! He had met the world on its own terms and had proved himself worthy. What matter that he was unique, the only one of his kind! Pride swelled up in the old one. What a glory was spread before him, and as he stood there he remembered, and then in his mind silently recited a portion from the liturgy.

"*The sands welcome each new signature into the memorabilia of Creation.*
Think how eternal you are, for you are part of creation.
God said: Let us make man in our image and the sun

Chapter 1 Attack on the Citadel

warmed that day. It was the same sun that warmed the day when you were born."

The meadow where the old one lived lay between two ridges and the rivulets that formed from the melting snow flowed to the west producing a stream that meandered through the meadow and then finally cascaded down a steep slope to a glen below.

Standing at the bottom of the glen on this warm clear spring day, the one-armed hunter looked up toward the tree line on the ridges surrounding the meadow and suddenly froze.

At long last there it was. In that instant he believed that he had finally found the creature he was hunting for many years. It had been thirty-six years since their last encounter, years during which the hunter had spent every spare moment searching the deep valleys and passes of the mountains of northern California for the beast that had maimed him. At first there had been just enough tantalizing bits of evidence to keep him going, but the number of strange sightings had increased during the last fifteen or so years. Then too, there were the reports of a peculiar track now and again and the story of a hiker or camper who thought that they had seen an unusual animal.

As he watched the creature high above the glen, the hunter acknowledged to himself that for all these years he had but a single purpose in life. He would find and destroy

Chapter 1 Attack on the Citadel

the being that had burst in on his life with animal fury. That creature had torn his family, his body, and his soul apart and there was no other course possible for him. In his strange quest, he, too, had bested the elements and, though not whole in body, felt that he was a match for any mountain and any beast on God's earth. The hunter had followed the reports of strange sightings carefully over the years and each time that he sensed that one might be about his creature he would try to track it down.

The hunter was a stocky man, about 5' 8" in height, his hair streaked with gray, his face tanned from long hours outdoors and with a week-old stubble beard. His left arm was missing below the elbow. He was well dressed for the cold and wore dark woolen trousers and a well insulated padded jacket. He was an excellent tracker and woodsman in spite of his handicap and usually could upon arriving at the area of the sighting, quickly identify the animal involved. Most of the time his search was fruitless for it was the bears which roamed the higher mountain elevations that were the cause for the original alarm. However, every now and again there was a strange track which he could not identify with any great confidence. These instances were what kept him going year after year. It was the report of such an unusual sighting by a cross-country ski party the day before that had drawn the hunter to the high sierra meadow just south of Donner Pass.

Chapter 1 Attack on the Citadel

The hunter, Roger Malek, now fifty-four years old had his first encounter with the creature in 1948. He had been eighteen years old when in the countryside in the foothills of the Sierra Nevada range east of Sacramento when the creature had bounded out of the woods and in a period of minutes on one summer evening, killed both his friend and his brother. As Roger had tried futilely to rescue his brother the creature had grasped him by his left arm and flung him with such force that the arm was mutilated beyond repair and subsequently had to be amputated at the elbow. In a single instant Roger had clearly seen the attacker in the moonlight and knew that this creature was not any animal native to the northwestern woods. There was no question in Roger's mind but that it was a massive erect ape-like creature that it had been their misfortune to meet that night, although he could never convince the authorities that this was so. This time, Roger had been tracking the animal through the glen and knew that the day old tracks he had followed were not made by any bear.

There could be no mistake he thought as he waited hidden in the brush scanning the hills above for some sign of movement. Then, there on the high rock ledges above, he saw it, an enormous creature black against the sky. It was standing erect and motionless on a rock promontory with its head and arms raised toward the sky.

Roger had waited a long time for this moment and knew there might be no second chance. He quickly aimed

Chapter 1 Attack on the Citadel

and fired several rounds in rapid succession and although the distance was very great Roger could see the creature go limp and fall. Gravely wounded the creature staggered to its feet, looked down at the glen for a moment and retreated about twenty feet toward a large rock outcrop where it collapsed in the soft packed snow.

Roger waited for a long while. Seeing that the animal was not moving and sensing the significance of the moment he made his way back down to his jeep to radio for assistance from the sheriff's office. He knew that he would need help with the massive carcass so that it could be brought out of the mountains. Now thought Roger, his story would finally be believed!

Roger knew the sheriff and each had great respect if not liking for the other, so it was with little conversation that Alan Garfield was informed of the possible capture of one of the legendary manlike creatures that were reputed to roam the western mountains. When Roger's call came in, the sheriff had just returned from the cemetery and viewing his wife's headstone for the first time.

Sheriff Garfield, a tall muscular clean-shaven middle-aged man in his fifties with distinguished looking graying sideburns, did not like cemeteries but in the past few months since his wife had been buried he had been a daily visitor. Some thought that it was out of a profound sense of loss, and indeed it was in part, but it was due more to the fact that he suddenly found that he had no place to go and

Chapter 1 Attack on the Citadel

nothing to do. He had never realized just how much he had relied on a single person to make his life full and complete. They had been married for thirty six years and although they had no children their lives had been quite full and rewarding. They had supplied each other with all of the comforts of a satisfactory human relationship and through the years they spent with each other neither had realized just how great the support that they had given to each other had been. Now, with his wife gone Alan did not like the thought of having to go home at the end of the day. He despaired arriving at a dark and empty house devoid of the familiar sights and smells, a place now with no warmth or cheer. He could not help wondering if this was the way it was going to be from then on.

Alan Garfield was the sheriff of a small town south of Lake Tahoe in the foothills of the sierras and had held that position for the past twenty of his thirty-seven years with the department. He had thought of retiring often in the past few years, but kept postponing any decision when he thought about how much he liked what he

did. It was not so much the job, but being with the friends that he made within the department, the regular routine of each weekday, and every now and then of being involved in a particularly interesting case. He was constantly being consulted since he knew his business and he also knew more about the history and problems in the area than anyone else on the force. In any event retirement

Chapter 1 Attack on the Citadel

was out of the question now. Aside from his wife, his home, and his work, Alan had only one other interest. He was fascinated by the stories of the strange manlike creatures that were reported to roam the wilderness areas of the northwestern states.

His interest in this subject had begun in 1948 when, as a young police officer he had visited the site of a double murder in which according to the lone survivor, Roger Malek, the perpetrator was a giant manlike beast. Unfortunately, the only evidence other than the witness's testimony in that case was several animal tracks at the scene of the crime. The tracks were unusual and large, but were not considered to be sufficient either to verify or to disprove the survivor's story. The crime had never been solved and thus it became just one more in a series of unexplained phenomena attributed to these mysterious manlike creatures.

Through the years, Alan had collected books and articles concerning these creatures and probably had the most extensive file available anywhere on the reported sightings and the associated folklore. However, as is the case with all folk tales, the stories that came from different sources often contradicted one another. Nonetheless, he was able to pull together some threads that ran through all of these reports and that seemed to make some sense.

They were described as massive, probably seven feet tall and weighing perhaps five or six hundred pounds. The

Chapter 1 Attack on the Citadel

creatures also walked on two legs and their bodies were covered with what was described as long black hair. These animals were reported to have been seen in places as far north as the Canadian tundra and to the south all through the southern portions of both the Rocky and Sierra mountain ranges.

Alan could not bring himself to believe this kind of an extended range for such a creature if it indeed existed. It was just too large and therefore if the beasts were real there would have had to be very many of them to account for the sightings. Alternatively, if their population was small, their long migrations within that extensive range should have caused them to confront mankind many more times than had been reported. In either case they would have long ago ceased to be the enigma that they were at present. Alan had concluded that if there were any truth to the legends at all, there were probably very few of these beasts and they maintained themselves within a very limited geographical range. Of course, he mused, they would have to be amazingly cunning to survive so near human civilizations without even one ever being caught.

Other than the single incident that Alan had investigated early in his career he had no other reason to believe that there really were such creatures. However, like all human beings he was always looking for and simultaneously fearful of finding some evidence of the unknown or the supernatural. It had been his hobby for a

Chapter 1 Attack on the Citadel

long time and now with his wife gone it had become a still more important part of his life.

Alan, excited by Roger's message but in complete control of the situation, quickly made arrangements for the dispatch of a police helicopter and then also notified the office of the county medical examiner just in case Roger was right and that he had brought down one of these elusive man-like creatures. Within thirty minutes the sheriff and his deputies along with an animal control department veterinarian were on their way to the high meadow where Roger waited.

In the meantime, Roger had retraced his steps back to the glen and began the ascent to the ledge above where he could still see the creature lying. Cautiously he approached the enormous hulk lying on its back until he was close enough to see what he had shot. As he looked at the creature he could see that two of his bullets had hit. One had torn through the left side of the creature's face just below the orbit of the eye and the other had smashed directly through the sternum into the chest cavity. The wet snow was mixed with blood and its pink rivulets were trickling down the slope toward the meadow below. The animal, big and obviously powerful, was surely some kind of an ape but not quite like any other ape he had ever seen.

Aside from its huge bulk and body covered with long hair, he thought it looked unnervingly human. Its face had only a relatively modest brow ridge and it had an unusually prominent jutting chin. It also had a clearly distinct protruding nose and robust lips. Roger thought to himself that it showed a remarkable resemblance to drawings he had seen of early manlike creatures. Its body was covered in a coarse long black fur except for the hair on the chest and on a short beard which were silver-grey in color. An enormous malformed ape perhaps, Roger thought, but at that moment he had no doubt that it was the same creature that had attacked him and the others that night long ago, crippling him in a terrible and painful moment.

After all of these years of searching, Roger was both frightened by the beast before him lying on the ground and yet exhilarated at the same time. As he timidly approached and examined the animal he noticed that the creature shuddered occasionally and that it was still alive although breathing very irregularly. While Roger would have liked to take this of all animals alive, he cocked his rifle so as to finish it off if it made any threatening move as a black shadow seemed to pass over him when he remembered its immense strength. The creature shuddered again and then Roger heard in a low guttural tone.

"Hunter," it gasped, and Roger instantly recoiled.

"Finish your work," the creature continued as a mixture of blood and frothing saliva trickled from its mouth.

"I will forgive you," it sighed "and you will do both me and all of mankind a great service."

Chapter 1 Attack on the Citadel

Roger looked at its face and into its eyes. It was looking back into his.

Roger, astounded, fell back in a state of shock so surprised that he did not even think to raise his rifle. In the next few moments as an element of reality and of composure returned, time seemed to stop and for Roger the thin and elusive boundary between what is human and what is not disappeared.

Roger continued to stare at the beast for some time until he was suddenly jarred to his senses by the sound of a helicopter landing just below in the open meadow. He once again looked at the creature and then quickly turned and made his way down toward the sheriff's party which had just arrived.

Roger descended from the ridge above to the meadow and approached the helicopter breathlessly, shouting and stumbling as he ran.

"Alan! Alan! - am I glad you're here!"

"Up there - I shot it! - It talks! - It talked to me!" Roger stammered.

Roger then looked around and seeing the rest of the party he took several deep breaths in order to collect his thoughts. With Sheriff Garfield were two deputies and a veterinarian with the state animal control department, all of whom were staring at him.

"It's like I told you, I shot this animal up there on the cliff," he said more slowly.

Chapter 1 Attack on the Citadel

"All right, Roger, settle down, that's why we're here. Is it still there?" asked the sheriff.

"Yes," replied Roger as he pointed toward the black form lying in the snow on the ridge above.

"I have to tell you that I'm very disappointed in you", said Alan, "As an experienced hunter you know that you broke the primary rule of hunters", he said disapprovingly"

"What's that," asked Roger.

"Know what it is that you are aiming at! - Why in hell did you shoot an animal without first clearly identifying it? What if is was another hunter?" said Alan, obviously annoyed with Roger.

"I knew exactly what it was," replied Roger quietly with conviction.

"I hope so" said Alan, "you may have been lucky this time. Let's go up and see just what it was that you shot".

"Wait a moment" said Roger, now somewhat composed.

Taking the sheriff aside as the others unpacked their cameras and other gear, Roger continued.

"Alan, before we go up there, I have to tell you what happened. After I called you I made my way up to the ridge above to get a better look at the animal that I had shot, and I got close enough to see it very clearly. It's surely an animal, but unlike any that I have ever seen before - and most important, it's still alive! At least it was a little while ago."

Chapter 1 Attack on the Citadel

"Are you sure?" said Alan, "We've been hoping for this to happen for a long time. Now maybe we can identify this thing and end all of this speculation about supernatural creatures from another world or another time."

"There's one more thing," said Roger excitedly. "When I got close enough to see where it had been hit, as I said before, I could swear that it talked to me!"

"Really, and what did you think it said?" asked Alan

"I think that I heard it say that I should put it out of its misery and that I would be forgiven for doing it", replied Roger.

"What I think is that you're a little crazy!" Alan responded, "maybe you've been up here in these mountains alone too long. What you probably have here is some kind of unfortunate deformed bear and a good dose of imagination."

"Everybody ready?" said the sheriff to the others, "Let's go up and have a look!"

The party made its way up the hillside. When at last they reached the rock ledge they all stood motionless for a moment and gazed at the crumpled hulk.

"You're right, Roger, that's no bear!" said the sheriff slowly as he and the others surveyed the creature.

"What is it?" asked Roger softly, as he watched the veterinarian make his way cautiously around the beast.

Chapter 1 Attack on the Citadel

"Well," the veterinarian said "I don't know exactly what it is, but it's definitely an animal, a primate of some sort. A gigantic monkey, an ape, you know."

The creatures' continued and irregular breathing proved that it was still alive but it now appeared to be in a comatose state and made no other movement.

"Give me my tranquilizer gun," said the veterinarian softly.

Wasting few motions, he quickly gave the beast an enormous shot of stunning agent.

"Get this thing chained up quickly", the sheriff instructed his deputies after some minutes, "and be damned careful!"

The sheriff also put his gun close to the creature's head but still unsure of what he would do if it threatened the others who were carefully shackling its arms and legs together. They then secured an additional chain around its neck, attaching it to the one around its feet, so that the head could not be brought forward.

"I guess that should do it," said the sheriff, "there's nothing more to do here except get some pictures. Let's get it back to the helicopter and tell the office what we're bringing in."

After they had quickly documented the scene, the five men dragged the beast down to the meadow and secured it in the helicopter rescue pod for transport.

Chapter 1 Attack on the Citadel

"There's too much weight for us to transport everybody this trip," said the veterinarian to the sheriff, "why don't you and Roger take the jeep back. In the meantime, we'll try to do what we can to keep this thing alive. It looks pretty bad, but what a find it will be if we can patch it up."

"OK," said the sheriff, "we'll see you back in town. Do the best you can but keep this whole thing as quiet as possible until I can get back to make my report and find out what protocol is in this case."

The veterinarian and sheriff's deputies climbed into the helicopter which lifted off a few moments later bearing its unusual cargo. Left alone in the meadow, Roger and Alan just looked at each other for a few moments.

"It's the one, I'm sure that it's the animal that killed my brother and my friend"

"Maybe Roger, that remains to be seen. Anyway, there are a few more hours of light. How about looking around to see if we can find anything that can help us figure this thing out? "

Chapter 2
The Cave

The two men started back up the ridge to the place where the creature had been shot. It was afternoon, and the day was still pleasant enough. Below, the meadow was peaceful and the air seemed to possess a special fragrance. On the ridge, as Roger and Alan stood for a moment on the same high rocky ledge overlooking the meadow catching their breath in deep gulps they could not help looking up toward the sky at the soft multicolored clouds drifting by seeming so close that they could almost touch them.

Although mostly obliterated around the scene of the shooting, there were clear tracks left by the creature as it had approached the ridge.

"Let's see if we can find where these tracks came from" said Alan, as the two of them silently began to follow the trail.

The snow was soft and there was little difficulty in following the trail along the side of the ridge. In about fifteen minutes they came to a place at the base of the cliffs where the track disappeared between two high rock columns and entered a rather protected place not unlike a courtyard where the cliff overhangs above kept out most of

Chapter 2 The Cave

the snow. At the far end of the sheltered area there was a narrow opening in the wall although there seemed to be some sort of wooden barrier across it. When they reached the cave, the two men examined the entrance and found that there was indeed a door. It was crudely made out of the trunks of small trees, each lashed securely to the other and showing the obvious signs of human craftsmanship. Although it had no hinges there were handles and the two men lifted and forced this heavy barrier aside so that they could enter the cave beyond. They judged the door's weight at around three hundred pounds.

Inside, the cave was dark but smelled quite fresh and herbal and in the dim light that filtered through the entranceway. Alan could make out a table with an oil lamp on it. Shrugging off his amazement at finding this remote habitation he proceeded to light the lamp and the two men looked about the shelter.

The cave was really a cozy apartment. There was a main room about fifteen by twenty feet with a roof that reached about ten feet overhead. The floor was dirt but level and dry. There was also a fireplace and several cracks in the rock roof above seemed to have served as natural ventilators. The furnishings were made as was the door, out of very crude wooden stock lashed with natural fibers. There was a table, a large chair, and a bed of similar proportions with a layer of heavy straw mat as its mattress. A few manufactured items were also visible - a flute that

Chapter 2 The Cave

seemed to be of silver and some other more mundane items such as a pair of reading glasses, and some dishes, pots, and pans. Some very large old trunks were also stacked at the far end of the room.

In a smaller connecting alcove a natural spring flowed down the side of the rock wall into a small elevated pool. In this room Alan and Roger were astonished to find what was obviously a latrine made of stone ingeniously engineered so that the spring's outflow continuously washed wastes through a rock cut to the outside of the ridge wall. A third alcove contained three large oil drums, two of them empty, and a number of crude wooden bins filled with dried berries, mushrooms, a variety of herbs and nuts, and sun-dried fish and meats.

"A simple and appropriate larder for anyone who tried to winter in this area," thought Alan.

After investigating and even tasting some of the provisions in the alcove, Roger and Alan made their way back to the main room and carefully opened one of the trunks. To their surprise it was filled with published books written on a wide range of subjects and in many languages. All told the men estimated that there were hundreds of such books in the cave.

One of the trunks, however, was somewhat different in content and contained handwritten treatises on a variety of medical, scientific, and legal subjects. All apparently written by a certain "M. Newman," these were far above

Chapter 2 The Cave

Roger's or Alan's level of comprehension as they leafed through them.

As they examined the manuscripts more closely, Alan noted that there was also a journal series entitled "Laboratory Notes: 1929-1968" by a Dr. Joseph Newman, D.V.M., and dated 1969.

"Well," said Alan, "I don't know what this all means, but it appears that this Dr. Newman used this place as a refuge for one reason or another. As for the creature, it's possible that it may just have wandered near the entrance of this place just as we did and the two findings may be completely unrelated, In any event, we'll send a crew in here tomorrow to pick up this stuff and try to find out about our Dr. Newman. I can't believe all the traffic we have up here now, hikers, campers, skiers, even recluses. Not too long ago, to be caught up here in the winter would have been certain death".

"What do you say we get back to civilization and find out about your monkey?"

Roger agreed, and as the two men set off down the ridge, past the meadow and down to the glen for the long ride back, both reflected on the day's events and although they did not talk about it, they were both apprehensive about what the next day might bring.

Chapter 3
Resurrection

The following day the helicopter party returned to the meadow to bring back Dr. Newman's belongings and to determine whether there might be signs of him or of any other "creatures" in the area. The night had been a fretful one for Alan, and he could not help but think that the cave and the creature were indeed related somehow.

The ape-like animal had been brought to a local hospital where during many hours of surgery through the late afternoon, attended by a group of human trauma specialists and assisted by several veterinary surgeons, its condition was stabilized before it was sent to the jail. At the prison special hospital cells were available that enabled the physicians to set up an intensive care unit under conditions that at the time were considered necessary for both the further treatment and confinement of the creature.

One of the bullets had penetrated its chest through the sternum, sending bone fragments throughout the chest cavity with several lodged near the heart. The bullet itself had been deflected by the sternum and exited on the right side of the body after piercing the right lung. There was massive internal and external bleeding and if it were not for the quick administration of human plasma just before

Chapter 3 Resurrection

takeoff, there would have been no chance for its survival at all.

The second bullet had entered just below the orbit of the eye and exited in the region just in front of the left ear, causing major damage to the upper jaw structure and traumatic damage to the left eye. This, the doctors thought, would probably result in its loss of sight if the creature were to survive at all. They had repositioned the shattered facial bones, removed the fragmented teeth, and pinned and wired the entire jaw complex together. Not unnoticed by the physicians was the fact that this ape had had contact with human medicine before - there were amalgam fillings in several of its molars.

To keep the ape from damaging itself when it awoke, it was placed on several mattresses and shackled to the floor in the jail hospital cell spread eagle so that it could only move its arms or legs a few inches toward its body. To make sure of its limited activity, it was also put on a heavy medication regime of both sedatives and tranquilizers. Food and drugs were administered intravenously, and the doctors agreed that the ape would have to be kept in this drugged condition for some time if there was to be any chance for its survival. Even so, the consensus was that there was little hope.

After the major surgery had been completed, the attending physicians and veterinarians had an opportunity to get some vital statistics. The creature measured six feet

Chapter 3 Resurrection

eleven inches from the sole of its foot to the top of its crown and weighed three hundred and eighty pounds, most of it distributed in the upper torso and concentrated in the massive chest and arm muscles. Coarse black hair covered the entire body including the genital area, which, when examined, showed that the creature was a male.

One of the veterinarians also observed that it did not have an opposable first toe on its foot, one of the traits that distinguish the apes from man. Apes have a prehensile foot with an opposable first toe while humans, even extinct species of man, do not. Also, he noted that while apes generally have a less erect stance than humans because of the difference in where the vertebral column meets the skull, and that an ape's canines are also larger in proportion to the other teeth, that these features were missing in the captured ape. Further, that in apes the cerebral portion of its skull is small relative to the size of the face and that older ape males have well developed supra-orbital bone structures that produce a striking and heavy brow structure. All of these additional typical ape characteristics appeared to be highly modified and reduced in this creature.

The creature's chest and facial hair were silver-grey in color and similar to that of older males of the larger ape species. Blood pressure at 110/80 appeared to be low, but may have been due to the recent trauma. The heart rate stabilized at sixty beats per minute and blood workup showed a type B antigen grouping with a typical primate

Chapter 3 Resurrection

blood cell distribution pattern. Serum enzymes and blood chemistry were generally similar and within the ranges found for both humans and other known primates. The doctors judged the ape to be between thirty and forty years old, although extensive arthritic condition of its bones and hands suggested that it was possibly much older.

While none of the veterinarians present could classify the creature precisely, they all agreed that it was clearly related to existing anthropoids thus eliminating consideration of its supernatural or extraterrestrial origin. Some ventured to guess that it was perhaps a circus freak generated as a hybrid between two species of great apes of the family *Pongidae,* but they soon concurred that while this was genetically possible they felt that such a mating should have been near impossible for behavioral reasons.

The most unusual aspect of this particular ape's features was its face. It had a peculiar light ruddy complexion and much higher forehead and a more reduced supra-orbital brow than any other known ape. It also had a distinctly elongated nose, rather than a typically truncated one with nasal passages opening almost on the face as was the case in most other anthropoid ape species. The lips, too, were peculiar in that they were full and the mouth relatively small. Also, the teeth were much smaller with less developed canines than would be expected in apes this size. Finally, the ears were relatively large and the chin was much more prominent than in any other apes. The doctors

Chapter 3 Resurrection

also noted one last thing, perhaps the strangest thing of all - the ape, unlike any ape seen before had light blue eyes. All in all, the face was extraordinary and while striking it was not an unpleasant visage based on human standards. The doctors estimated that its cranial cavity was about 2,400 cubic centimeters, much larger than that of any existing primate including man.

During that same day, the sheriff filed his report, including all of the previous day's details with one exception. He did not mention Roger's excited garbled statement blurted out during that first agitated moment when the helicopter party arrived, or his subsequent statement to him. After it was confirmed that the creature was in fact some type of ape, the story of the capture was released and during the furor that followed every form of speculation was voiced as to its origin. Because it was considered quite unreasonable that this creature or a group of them had lived in the area for hundreds of years without one being captured before, the general consensus was that this was an isolated incident and that the animal had recently escaped from some research institute or circus. There was only Roger's testimony to place the creature in the general area at least thirty-six years before, and that incident had occurred about thirty miles east of Sacramento, not in the high sierra country.

Chapter 3 Resurrection

A cast had been made of the creature's foot and the authorities planned to compare this to the cast of a footprint taken in evidence at the time of Roger's brother's murder. It would be several days however before the evidence would be available for comparison.

Roger was now agitated and worried about whether they were indeed the same. He was especially worried now because he still believed that the beast had talked to him! A sense of dread began to overwhelm Roger who knew that more had happened on the night of the murders - things he had never dared to tell anyone, but that this creature might now reveal.

Chapter 4
Genesis

The following day was a busy one for Alan. As soon as he arrived at the office he dispatched a detail to recover the trunks and other items of interest that had been found in the cave. It took several trips and not until evening were the belongings of "Dr. Newman" received, properly marked, and placed in the jail exhibits storage area for safe keeping. Meanwhile a rapid search through the Sacramento office of records yielded a list of local residents named "Newman" and whose first names began with the initials "J" or "M". Unsurprisingly, there were almost fifty names on the list, and all of these, Alan realized with some apprehension, would have to be contacted.

The cave, together with its mysterious contents was a curiosity in itself, but the possibility that it was somehow related to the presence of the creature in that vicinity made it even more fascinating.

Perhaps, thought Alan, Dr. Newman was the ape's owner and had taken advantage of the remoteness of the area to carry on some special scientific studies. It was plausible and would not, he knew, have been the first instance of primate research conducted under natural and remote conditions. Alan recalled reading of many such

Chapter 4 Genesis

studies recently carried out in Africa where scientists actually joined baboon and gorilla troops for extended periods in order to study their behavioral characteristics. Indeed, some scientists had actually adopted young anthropoid apes into their families and for several years raised them with their own children. Uneasily, Alan thought how these studies had also shown that many of these apes were highly intelligent, and when taught appropriate sign language, could communicate with their trainers under controlled conditions. Late that afternoon, one more development with regard to the creature had come to light. When the cast made at the scene of the Malek murders many years before was compared with the one taken of the foot of the creature that lay drugged in the hospital cellblock, it appeared that they were very similar, perhaps identical! But, he mused, where had it been for the last thirty five years?"

It had been a long and harrowing two days. At dusk, Alan, pondering these most unusual events, left the office and went straight home. He retired early that evening, glad to close his eyes and forget about the ape at least for awhile.

Early the next morning Alan made some inquiries and then placed a call to Dr. Paula Chapman, head of the anthropology department of the University of California. The books in the trunks had to be inventoried anyway and he felt that if the texts and journals were related to primate

Chapter 4 Genesis

development, as they well might be, it would be best to have someone technically competent to do the job. Paula, of course, had heard of the recent events and was all too happy to have the chance to observe the creature first-hand and to examine the material found at the capture site.

Paula, at forty-two, petite but well proportioned and with slightly graying short curly black hair, was one of the youngest department heads at the University. She was a dedicated scientist who had chosen a career rather than the rewards of family life. This had happened, at least in part because her anthropological studies ever since her student days had kept her away from home traveling in Africa and the islands of the Pacific. Her special area of interest was early man and her work centered on the social and physiological development and attributes of the earliest species that had been conserved and ultimately evolved into modern man, *Homo sapiens*. Over the years, Paula had developed a theory that many or all of the earliest species had not perished from the earth without a trace but had interbred with more advanced species, thus introducing their distinctive genetic material into developing mankind. As a result, she had predicted that from time to time a human being would be born with a combination of so many of the more primitive genes that the person could be considered a "throwback" or living representative of one of those extinct groups of manlike creatures.

Chapter 4 Genesis

Paula's research had involved a search for these rare beings throughout the world. During the past twenty years she had acquired a file of about a hundred case histories that met some of the criteria she had established and she was preparing a monograph on the subject. However, for all of her hopes and efforts, she had still not seen what she believed might be an unusual birth of a living human that truly represented an earlier species of man. Paula rearranged her schedule so that she could get away from her university duties for a few days and arrived at the county jail shortly after noon.

"Well, thank you for coming, Dr. Chapman. I hoped that you would not mind this imposition."

"No imposition at all, sheriff, I meant to contact you anyway so that I could see this animal first hand – call me Paula".

"I'm Alan – I'm glad you're here to help us."

"I'm sorry to rush you, but let's go upstairs to the infirmary. I just had a report on the ape's condition. It appears to be stabilized now and the prognosis indicates a good chance for survival."

The two went up to the hospital cell where they observed the creature stretched out on the floor still tethered and under heavy sedation.

"The doctors say that if it survives for a few more days, they can pull out the plumbing and transfer it to the ape

Chapter 4 Genesis

compound at the zoo. They feel that these bars may not hold it if it has a mind to get away."

He paused, and then went on.

"Incidentally, we also have another problem. This beast appears to have been responsible for killing two people about thirty-five years ago. In fact, the reason it's here is because a third man present at that encounter named Roger Malek tracked it down and shot it. We thought that he was a bit nutty all these years but it turns out that he may have been right. It seems to me that if this proves to be the case, the state will have to destroy it just as it does in all of these instances involving killer animals".

Paula had been observing the creature as Alan talked, and could not help noticing how striking and sensitive looking the facial features were, and how tall and straight its body was. The word that kept entering her mind was "throwback." This, she thought might not be an ape at all, but perhaps a human offspring that really bridged the narrow gap between the anthropoid apes and man!

"Let's go down and have a look at the documents," Paula said abruptly, and the two of them left the sleeping creature to his dreams.

Downstairs, in the exhibits area, Paula and Alan looked over the trunks that had been brought out of the mountain cave.

Chapter 4 Genesis

"There are ten in all just as we found them, nine filled with what appear to be published scientific works, and that last one over there, filled with handwritten manuscripts."

"How do you want to handle this job?" asked Alan.

Paula replied, "Why don't you let me browse through the material for a little while and after I have some better idea of what's here, we can decide what should be done."

"Good," responded Alan. "If you need anything, just ask the exhibits storage clerk for what you want. There's a small office over there that's used for reviewing evidence. It's not too fashionable but it's got some comfortable chairs and good light. I'll stop back and see you later."

After Alan left, Paula put her things in the little room and returned to the storage area.

The trunks were black metal, all of the same make and not very ancient although they were dusty and stained especially where they had rested on the dirt floor. She opened several to get some idea of their contents. Each was lined with a heavy plastic bag in which were packed volume after volume of printed works. Scanning the titles quickly, Paula noted that they covered almost every field of human knowledge – music, history, mathematics, literature, and economics, to name a few. Not only were these works on a variety of subjects but most were advanced treatises not usually found together. True, a particular scholar might accumulate a series in his special area of expertise, but to find so many subjects covered at this level of sophistication

Chapter 4 Genesis

was incredible. No individual could be master of all of these subjects, she thought.

None of the texts checked were newer than 1969 nor were any older than 1960. It seemed that the books had been collected in 1968 or 1969 and if this was true, and considering that scientific texts only have a useful lifespan of about five years, the collection apparently represented an attempt to obtain the most recent knowledge available at that time. In each of their varying subject areas the books seemed to be the most recent, regardless even of the language in which they were published. Moreover, all of the books that Paula looked at had also been read – as evidenced by penciled comments in the margins of many pages.

Paula then turned her attention to the tenth trunk. It was like the others but was filled with handwritten journals covering many of the same topics she had seen among the other books. The handwritten works seemed to be distillations, reviews, or extensions of those published books but for the most part Paula was unable because of her limited expertise, to evaluate the journals. All of these manuscripts carried the name "M. Newman" on the flyleaf.

Under several layers of these manuscripts Paula came across a series of five journals labeled "Laboratory notes: 1929-1968" apparently written by a Dr. Joseph Newman, D.V.M. A sixth volume in this series was dated "1969."

Chapter 4 Genesis

Paula brought these into the office and opened and began to read the most recent of the series.

"Everyone at one time or another sits back and takes inventory of his life and what it has meant, both to him self and to the world in which he lives. My own life has been rather unusual, to say the least, but it is only lately upon reflection in these past few months that its unique aspects have made it seem both so absurdly humorous and so tragic that I must share it with someone. Of course, I still cannot tell anyone and so I am writing this story, to be kept by my son and used when, if, and as he wishes. I am now almost sixty-six years old. I have a fine son who has no legal existence and whose mother I never married, cared for, spoke with, or even slept with for that matter.

For two decades I was legally married to another lovely woman with whom I also never slept, and am recently widowed. My wife had three sons whose legal father I am, but who were in reality my grandchildren. The fact of the matter is that my legal wife was also my son's devoted mate, whom we all admired and loved dearly. It is also a fact and now quite amusing that although a grandfather three times, my marriage was never consummated.

Another incontrovertible fact is that my heart has weakened considerably in the past few years and I am soon going to die. I am not afraid of death, but I have a great concern for what will become of my son when I am gone. Many of the hard and cold facts concerning my son's birth

Chapter 4 Genesis

and development can be found in my laboratory notes, which are also entrusted to him. However, these notes do not explain anything about the circumstances that led to this tragic experiment or our subsequent feelings, nor the reasons for certain of our actions through the years. It is this history that I must relate and I hope, should these documents be made public, that the reader will be compassionate and find it in his heart not to judge me too quickly or harshly for what I have done. I suppose that it is best to start at the very beginning.

I was born in 1903 in San Francisco and named Joseph. My parents were second-generation Californians, whose ancestors had left Europe in order to escape religious persecution. They owned some property that included a small dairy farm on the peninsula just south of the city. My childhood was not extraordinary in any way. I was a rather introverted boy, somewhat shy, and at an early age knew that I would like to work with animals as a livelihood. Since I had done particularly well in the lower schools, by the time I was sixteen which was just after the end of the World War, I entered the university where it was my intention to become a veterinarian. My studies occupied most of my time and I was quite successful at my chosen discipline, which resulted in my receiving the degree of Doctor of Veterinary Medicine, in the year 1927, at the age of twenty-four.

Chapter 4 Genesis

Although I did not have much of a social life or follow the world's events to any great degree during this period, I must say that I was particularly distressed at the Scopes "monkey trial" case that took place in 1925. A young teacher, John Scopes, has been accused of teaching the theory of evolution according to Darwin and others in the public schools. The ridiculous theme presented by the prosecutor was that the evolutionary lines for man and ape were different and that the heresy of teaching that they were not could not be tolerated. The defending lawyer based his defense on science and logic. However, I could not help but think how the absurd nature of the arguments would be illuminated by the presence in that very courtroom of a witness, a living example of an early apelike man to plead his own case. A fantasy perhaps, but I mention it here because it is in fantasies such as these in which we indulge that the seeds of deeds to come are planted. In this case, that fantasy was the first act in the tragic history that was to unfold.

In 1928, I obtained an appointment at the San Francisco Zoo as staff veterinarian, a post which occupied about half of my time. I also started a private practice on the peninsula in the town of San Mateo and proceeded to establish myself as a reputable member of my chosen profession. The work at the zoo was interesting and since I was the only veterinarian, there was a wonderful opportunity for me to work with a variety of unusual creatures. The most

Chapter 4 Genesis

interesting animal there for me, however, was a large male mountain gorilla that, through his strength and innate intelligence, lorded it over the other zoo inhabitants.

One day, I received an urgent call from the zoo authorities to report to a circus troupe performing in San Francisco and take possession on behalf of the zoological society, of a seriously ill female gorilla. When I arrived, I found the animal weak and feverish and with little wonder considering the cramped quarters and poor diet the animal was subjected to when not performing. The society had purchased the animal for a nominal amount and considered the gamble worthwhile if it could be brought back to health and mated with their male. I treated the beast, but it was several months before the female was restored to health.

Upon reporting the good news to the society, I was then also charged with making suitable arrangements for the mating. During this period the female had grown quite accustomed to the male, and one might say almost appeared fond of him. At any rate it became quite obvious that the female had once again resumed her normal cycle and was now in a condition to accept the male.

At this point I must digress for a moment and try to explain the basis for my subsequent actions. After all of these years it is difficult to remember – or maybe I don't want to remember – the elements that consumed me and made me believe that what was terribly wrong was really right. Perhaps I need only say in explanation that all of us have things about us that we are so ashamed of having done that we dare not tell anyone – not our dearest friends – our brothers, wives or even our priests! Things that were so stupid, so out of character, so thoughtless and unmindful of the consequences that we don't even like to remind ourselves of the matter and drive the secret into our subconscious.

Mine was one such deed. For a period of twenty-four hours, I became obsessed with the idea of the possibility of producing a human-ape hybrid. Biologically, there was a good chance that it could work. Both groups belonged to the same superfamily, *Hominoidea*. The chromosome pictures were also similar. We shared blood-group antigens and certainly anatomical characteristics. The fact that the gorilla and human belonged to somewhat different animal groups was not a significant obstacle since other such inter-group and unusual crosses between related creatures had been made in the past. The size difference between the species made for a physical barrier which meant that the only viable cross could be the insemination of a female of the larger species in order to have space to contain the fetus. The behavioral problems could be overcome by using artificial insemination, a process already common in animal husbandry. What if such a creature could be produced! What a boon to science and to our understanding of the nature of man I thought. What a prize piece of evidence to present at the next "monkey trial"!

Chapter 4 Genesis

My actions from that moment on were purely technical and mechanical. I committed myself to the deed with no further thought of the possible repercussions and, almost as an automaton, proceeded that evening to prepare the female gorilla's food with a heavy dose of sedative. I suppose that in the back of my mind I thought that I could stop at any step along the way just as an alcoholic sees himself stopping after each drink, or a gambler after each wager. Later that evening I entered the cage with the glass insemination apparatus and within moments completed the procedure. The entire process from beginning to end took no more than ten minutes. I was the human donor in this experiment. Later that night as the full meaning of what I had done sank in I broke into a cold sweat. What if I had been right?

What if I had created a new individual, neither animal nor human but a person, a part of myself, a primitive but reasoning entity that would not be accepted by any community, a monster of my own flesh and blood? The next days and weeks passed slowly and since there was no overt result of my actions, I felt a little better. It had been a stupid thing to do and I would hide it from others and myself for the rest of my life and that would be the end of it, or so I thought. After six weeks, the female gorilla started to behave differently and seemed antagonistic toward the male in the next cage. She also no longer exhibited any sign of being ready to accept the male. Some

Chapter 4 Genesis

other slight differences in the way she ate, her body form and curious nesting behavior made me aware that the worst had happened. She had indeed conceived! The weeks and months dragged on and I began to pray that the offspring would not be viable and would abort naturally, but this also was not to be the case. At the end of seven months, in the year 1930 and in the midst of the great stock market crash and bank panic, my son was born.

Although I could not know it at the time, two other children were born in this same year that would profoundly influence my sons and my own future life. One was a little girl by the name of Mary White. The other was the second son of Benjamin Malek, a boy named Roger.

My own son was kept hidden by his mother most of the time but eventually I was able to get close enough to see him. At the outset I must say that to the untrained eye he was a gorilla for which I was profoundly grateful. He was covered with long black hair and although his relative dimensions were gorilla-like, he certainly would have had to be classified as the runt of the litter. There were several noticeable exceptions, however. First, his feet lacked the opposable first toe common to anthropoid apes and secondly, his cranium was much larger than that expected in a gorilla baby. The face however was the most interesting. His eyes were blue and he had an extended nose bridge and well defined lips. To anyone else examining the baby, they would have considered it a

Chapter 4 Genesis

grotesque and deformed gorilla. However, I knew better! Since my immediate anxieties were over, I settled back to watch the development of my son and at that time started this series of journals which you have before you. In order to identify him, I gave him a proper name befitting his parentage. I called him Mangor, Mangor Newman."

Paula put down the journal, closed her eyes and sat quietly for a long while. Outside it was getting dark and although she had not eaten any lunch, she was still not hungry. Now it had all come together. This was not the first time she had heard of human-ape intercourse. Primitive folklore abounded in such stories. Many African tribes had used apes for god-worship and in some cases the climax required union between man and ape. Similar stories came out of the far corners of the world wherever a suitable anthropoid ape co-inhabited an area with the human species. But no one before had ever considered the possibility of the production of a viable offspring! However, there were the stories passed down through the ages of grotesque man-like creatures that roamed the edges of civilization in the wilderness areas. Were these the heritage of mans folly thought Paula. Were there ever enough of them at any given place and time to produce viable and sustaining colonies that might have been evolving along separate but parallel lines to both human and their ape ancestors? Had not this same type of activity been going on for thousands of years before modern man arrived

Chapter 4 Genesis

on the scene? Could and did Neanderthal man cross with the anthropoid apes 100,000 years ago? If so, just what input might these viable crosses have had on the genetic makeup of modern man? The possibilities were endless, mused Paula, as she opened her eyes and called Alan.

"Hi, this is Paula. Can you join me right away? Also, can you get us some coffee and sandwiches? I have every intention of spending the night here and I think that you will also do so after you learn what I've found out about our friend sleeping upstairs."

Alan arrived within a few minutes with a pot of coffee and poured two cups.

"The food is on the way. Now, what have you found out?

"Well, for one thing," Paula began, "the creature and the contents of the cave are definitely related. There are a series of lab notes and Dr. Newman's handwritten journal that describe the creature perfectly."

"That's great" said Alan, "then we know who owns the ape. I'll have my people try to get in touch with him in the morning."

"There's more!" said Paula.

"What's that?" replied Alan.

"First, I think that Dr. Newman has been dead for the past fifteen years. I believe he died in 1969 at the age of sixty-six."

Chapter 4 Genesis

"Well, that's a problem isn't it" Alan mused." Wait a minute, who has been taking care of the ape all of this time? Surely, he couldn't have survived on his own for that long in that country. These apes don't hibernate do they?"

"No," replied Paula, "but there's still more."

"What is it?" asked Alan.

"Dr. Newman may have been a nut, but if we are to believe what I've read so far, and I think that I do, it seems that Dr. Newman was that creature's father!"

"What!" exclaimed Alan in a state of disbelief?

"The creature's given name is Mangor and he is a human hybrid formed out of Dr. Newman's artificial insemination of a female mountain gorilla. The birth occurred in 1930, which makes the creature fifty-four years old – a little older than our estimates, but about right."

At that moment the sandwiches arrived. After setting them out on a small tray Alan picked one up, sat down quietly and ate it while gazing out of the window at the night shadows. Paula was hungry and had her supper in silence.

After awhile Alan asked,

"Where do we go from here?"

"Well, I think that if we are going to be sensible about this thing tomorrow, that we had both best review Dr. Newman's journals tonight. Do you have to call anyone?"

"No," replied Alan slowly, "not anymore."

Chapter 4 Genesis

"Neither do I" said Paula, and with that they settled down to continue the history of Mangor Newman as presented by Dr. Newman.

They took turns reading the journal entries aloud.

"Mangor developed rapidly and after three months he began to explore his environment more fully and spent more and more time away from his mother's side. He was a delightful child, inquisitive and playful, and although he was somewhat stunted and thin for a baby gorilla, I knew him to be robust enough for what he was. We both came to look forward to the time each day that I spent at the cage and when I was even a little late, he would become quite agitated. Of course, maybe it was the treats that I brought to him – but I liked to think that he cared something for me also. When he was about seven months old, an incident occurred that made it clear that I was going to have to pay for what I had done in one way or another. Each day that I came to visit, I had greeted my son by name, saying;

"How are you today Mangor?" or "Hello Mangor, your father is here."

We spoke, or rather I talked to him in the way any father would talk to his offspring. One day however, after I greeted him with "hello Mangor" he responded in a low tone, and almost inaudible.

"Fa-a-a-ther"

It became instantly apparent, even though I had heard him babbling before and should have considered the

Chapter 4 Genesis

possibility, that an additional feature which he had inherited from me was the capacity for speech. At that moment, I knew that I would have to remove him from the public eye for my own sake. The alternatives open were to destroy my son of whom I had grown quite fond, or to take him away and somehow explain his disappearance. I chose the latter although now that I think of it, there never was any real choice.

First, I arranged for a special enclosure to be built in the basement of the building in which I carried on my private practice. It was a chance, but I felt that I could get away with it. One evening I repeated the sedative treatment that I successfully had used before and once again entered the gorilla cage and quickly removed my son. He made no cry and came with me quite willingly.

The next day I prepared a report indicating that the young and weak gorilla had died of natural causes and that I had properly disposed of the remains by cremation shortly after the autopsy had been completed. I'm not sure that my ploy worked for there were many questions asked about the nature of the illness, the speed of my disposing of the remains and my failure to notify the Society's officers before performing the autopsy. In any event, I was notified the following week that "because of the need to limit expenses during this period of monetary instability," my services were no longer required.

Chapter 4 Genesis

All things considered, I was glad to be let off so easily. I had my son to take care of now and wanted to get away from the area where I had both family and friends. I didn't know exactly what I was going to do but I was pretty sure that whatever it was I would have to do it alone. With that basic decision made I sold my office furniture and small practice for what I could get, packed my remaining goods and my son into my car and headed east of the Sacramento area. There was plenty of room there in the foothills of the mountains and plenty of livestock to care for so that I could continue to make a living and support my new family. At that time I envisioned my life as still being pretty normal and free but with the minor inconvenience of having to support a child-like and primitive creature, perhaps on a par with a childlike human being. Little did I know how wrong I was".

Paula set down the journal and prepared another cup of coffee.

"Do you want one?"

"Yes, that will be good," Alan replied.

"How are you doing?" asked Paula.

"I'm just fine. I didn't believe Roger at the time, but he was apparently right and the "creature" could talk."

"What do you mean?" said Paula.

"Well, when he first met us in the high meadow, he was babbling – at least I thought that he was. He was so obviously distraught and who wouldn't be expected to be

48

Chapter 4 Genesis

so after running into this creature. He told me that the beast had talked to him!"

"What did it say?" Paula responded.

"According to Roger, the creature had asked him to kill it and put it out of his misery! I didn't believe it and chalked it up to the result of the day's strain and a vivid imagination. I didn't even put it in my report!"

"I think that report may need to be amended," said Paula as she picked up Dr. Newman's Journal once again.

Chapter 4 Genesis

Chapter 5
The Early Years

Paula went on reading from the journal.

"The year 1931 turned out to be a rather dismal time with the country in the midst of a severe depression following the market crash of 1929. As a result of these conditions, I was able to secure a small house and some land about thirty miles east of Sacramento within my means. It was quite isolated and surrounded by low mountains, the foothills of the sierras that rose to great heights further to the east. Being so many miles from the city line in those days meant isolation, but this, for obvious reasons, suited me just fine. I limited my practice to the farm communities within a radius of about fifty miles and thus was never away from home for more than a day at a time. While work was generally hard to get, I fared pretty well considering all things. One of my first chores was to set up suitable living quarters for my son and to see to it that all appeared quite ordinary. It was not unusual for a veterinarian to have an animal holding compound associated with his office and so I built a fenced-in area adjoining the back of the house with access through the back door. The compound had some trees and large rocks

Chapter 5 The Early Years

in it and a proper shelter so that he could sleep during my absence in relative safety.

Whenever I was home, and each night, he stayed in the house and slept in a small bed in my room. He took his meals with me

and it was not long before I began to realize that he was undergoing every behavioral pattern expected of a human child, - only at a much earlier chronological age! He was toilet trained at ten months of age, quickly developed suitable table manners and mastered many of the games and toys made for three to five year olds. His first words were not his last! By one year, there was nothing I said that he did not understand and he spoke well and clearly with a vocabulary of about a hundred words. The fact of our isolation made us spend most of our time together talking, and he proved to be a constant source of questions which I was sometimes hard pressed to answer.

Early in the game and quite matter-of-factly, I explained who he was and who his parents were. There was no question in his mind that this was all right and proper and indeed for the rest of his life there was never any misunderstanding regarding this matter. By the time he had reached the age of two, I could give him explicit instructions with regard to how he was to spend his day in my absence and he would carry them out faithfully.

During his third year, I began a program of informal education to teach him to read so that his time alone could

Chapter 5 The Early Years

be more interesting. He responded quickly and by the time he was four, he could read and enjoy the children's stories and fairy-tales usually associated with the fourth and fifth grade levels. By this time he had grown considerably as will be seen in my laboratory notes and by the age of five, he stood about four feet tall, very erect, and weighed about fifty pounds. It was obvious that his full stature and weight were going to be much greater than that of a typical human, but probably near the upper range of the human growth-weight distribution curve. At least I hoped that he would not develop more like his maternal parent, considering that a male mountain gorilla could attain a height of near seven feet and a weight of seven hundred pounds. I might also add that I was very pleased that my son had an extremely pleasant and sweet disposition.

Many times I considered telling the scientific community about my son and the enormous scientific breakthrough that he represented. What new insights into the position of man in creation would be pondered and learned! What a great advance to our studies of evolution and anthropology, and with no less impact on the fields of physiology, medicine and psychology. My reluctance to disclose my secret at first was due solely to the personal shame that I associated with my actions, but as the years went by and I grew to love my son, with his quick smile, gentle disposition, and great intelligence, I could not even consider the possibility of his being treated as a spectacle or

Chapter 5 The Early Years

laboratory animal. Indeed, what would become of him? Would I have any rights in this matter? Would he be considered mine, either by family or property, or would he be considered the property of the Zoological Society? Thus, by the time he was five, and probably long before that, the question of disclosure during his or my lifetime was no longer considered. I knew that for the rest of our lives together that all others must be excluded and that any satisfaction with life would have to be of our own making. At the age of thirty-two and not knowing what my son's life span might be, this was a hard fact for me to accept. However, when I thought about it, I also understood that if he outlived me, his existence would probably be discovered.

We spent a good deal of our time walking in the woods and hills that surrounded our home. My son was agile and inquisitive and kept me on my toes with his constant questions that needed answers. It was when he was about six that I began to fully realize that there was something quite peculiar about my son's learning processes. I had suspected it for some time, but could not quite put my finger on it – the way that he associated distant facts and came to conclusions far ahead of me, the way he would grasp an idea, a thought, with the barest explanation, the way he would combine some common bits of information and present an idea with a completely new perspective for me.

Chapter 5 The Early Years

What I found was that I was no longer teaching him things, but that many times our roles were reversed and I was the student, not he. What I had considered to be a case of precocious development and which was probably due to the known differences in the rates at which anthropoid apes and humans grow and mature was decidedly incorrect. I had long realized that he would have an intellect the equivalent of that of a human being, but somehow, the combination of his inherited human brain patterns and intellectual capacity, with the more rapid rate of ape neurological and physiological development, had produced something unique.

What my son exhibited was a case of "hybrid vigor" in his intellectual capacity. This is a common enough phenomenon in which the offspring of two organisms exhibit characteristics common to neither parent. It may be bigger, hardier, more vigorous, more productive or in a wide variety of ways better than either parent. Such was apparently the case with my own dear son and I knew from that moment on that he would continue to develop intellectually at a rapid rate and not only achieve a level of normal human intellect early, but eventually surpass it and continue from there to what – brilliance? – Or perhaps something never before seen and measured throughout the recorded history of humanity! My life was truly exciting and at that moment, for the first time in any years, I felt pleased and satisfied. No longer was my son a dark secret

Chapter 5 The Early Years

to be kept in a closet, but truly a gem – surely one to be kept guarded, but to be displayed in the home and watched and enjoyed more and more with time. In the person of my son it appeared that the evolution of apes to mankind, and perhaps beyond, had progressed some seven million years in a relative instant of time.

In 1937 there were ominous signs in Europe, the harbingers of World War II which was to come soon. The next four years were to see its beginning and in California, the incarceration of a resident hominoid race by others. However, the effects of the long depression did end and I was able to earn enough to support my son's special needs. These were not ordinary needs although they were material in nature. We had tried clothes at one time, but found very quickly that they were an unnecessary encumbrance. The black hair that covered his body was some two inches long and served him adequately. I must point out that this was another unusual characteristic since neither parent grew such a long and silky coat. The gorilla's was short and of course man had long ago shed his. He also required no shoes.

As for his diet, he was omnivorous as are humans, and we cooked and shared the standard meals partaken in every kitchen in this country. Of course, my son's table manners were quite correct and he was a big help in the kitchen and generally keeping the house in order. We maintained the enclosure outside and he had made it into a delightful and

Chapter 5 The Early Years

pleasant garden spot where he liked to sun himself. In the house, he had his own room now with ordinary furnishings common to a young boy's habitat. He was also fastidious with his person, bathed regularly, and brushed his teeth each morning and evening. It had been a long time since I had even noticed that he was not like everyone else; he did not seem unusual or ugly to me. He was merely my son, healthy, happy and a pleasure to be with each day.

The material things that he did require, as I mentioned, were books. At first, I used the city libraries to obtain books for him, but as he began to read the major authors and wanted to come back to certain passages from time to time, this became too difficult and I started our own library. By the time he was eleven, he had a library of several hundred books covering all areas of human endeavor including art, literature, history, the sciences, politics, religion and philosophy. He was truly a prodigy, having at that time a common vocabulary in excess of two thousand words and the ability to reason on the most abstract levels. He had also by this age matured physically as well and approached six feet in height and a weight of about two hundred pounds, far out distancing his father in all respects. Even more exciting was that there was no evidence of any slowdown in his intellectual or physical growth!

He was trim and powerful and spent much time in the woods and the high places, which he liked especially well. We still had no neighbors, and he made every effort to

Chapter 5 The Early Years

avoid human contact when on his own. Once in a while he was sure that he was seen at a distance, but no one must have ever considered it anything more than the catching of a glimpse of an uncommon animal. No reports or searches ever resulted from these contacts. My son and I had settled into a pattern of life that, while unusual, was agreeable to both of us, rewarding, and probably no stranger than relationships other human beings had established at the fringes of society all through the ages in order to care for their special needs."

Chapter 6 An Extraordinary person

Chapter 6
An Extraordinary Person

During the next seven years, my son's intellectual development continued without abating. I purchased more and more books on every subject imaginable and in fact some of the more advanced treatises and the copies of ancient works had to be sent from various parts of the world. This was not an easy task since for most of this period the world was involved in a war of enormous magnitude that included every continent and saw the burning of the most cherished and priceless of the world's manuscripts and books. It was amusing to reflect, if indeed anything that came out of that time could be considered amusing, that at the same time human beings were wreaking such havoc and destruction, my less-than human son was acquiring, reading, and safeguarding those precious works!

There was no longer anything I could teach my son, and I could only serve by providing him with the materials he required for learning. Everything now was self-taught. He became proficient in eight languages in addition to his native English and he read texts in the languages in which they were originally written so as to gain the full meaning

Chapter 6 An Extraordinary person

of their authors. No subject matter was excluded and in each subject he progressed to the point where I felt he was on a par with our greatest scholars. At this same time he began to take an interest in everyday occurrences in society, reading several newspapers daily and becoming absorbed in the radio, which he valued as a tool in communication. He seemed to be turning outwards toward the world which he already knew very well, perhaps better than anyone else.

It was during this same period that he developed his musical skills and became proficient at several instruments - the cello, French horn, and his favorite, the flute. All of these he taught to himself and then played with excellent skill as he mastered each one. He began to paint, and his canvases were sensitive and delightful, displaying color and form executed - in interesting ways such as I had never before seen. He signed them M. Newman, and to cope with the library costs, we agreed that I might sell some of them in San Francisco. The paintings were highly regarded but it was I, unfortunately, who became quite famous as the artist. Indeed, some of these works now hang in an impressive number of museums around the world.

In 1946, at the age of sixteen, my son entered into the final phase of his development which was the natural extension of his years of acquiring knowledge. At that time he began to synthesize new information. In other words, he advanced broadly into new areas of knowledge by using the

Chapter 6 An Extraordinary person

most advanced information available as a spring-board. It was at this time that he began his journals, which he has continued up to the present. Perhaps it was because my son was an outsider that he seemed to see things more clearly. His visions of the way things should be were sharp, much sharper than my own and these things he wrote about in his journals. For instance, he developed a new theory of compensation and trade based on what he called the "caloric standard." Basically, this system would tie currency to the caloric value required by the average individual each day. Thus, for example, the dollar would be valued at three thousand calories, a value that would never be allowed to fluctuate, and so anyone who had a dollar would always have the price of a day's sustenance. This would eliminate inflation and possible loss of value. Since labor is the base of all wealth each dollar then would be translated into daily productivity and the only way to obtain more dollars would be by the making of real goods. The system proposed would have all of the benefits of the gold standard system and the additional benefit of an absolute standard. After all, people can suddenly decide that gold has no value and is no longer desirable. However, there will never be a question about the desirability of rice or wheat or corn. Imagine a loaf of bread always at twenty cents or a steak at fifty cents per pound.

In his journals he also dealt with the decline and destruction of the nation's cities; with the basis of racial

Chapter 6 An Extraordinary person

conflicts; with war and its inhumanity and the world's dwindling resources. He thought about these things and he formulated answers, in many cases quite novel, but which, I believe, were capable of solving many of the world's problems. He understood the potential bounty of the seas and outlined methods by which mankind could benefit from new drugs that could be obtained from marine plants and animals. He also devised a plan by which the fouling organisms of the sea, long a bane of shipping and seaport activities could be made to provide an unbelievable bounty as a food material. In fact, he had calculated that a "fouling" farm only fifty miles on each side could feed the entire population of the United States. What a boon that would be to the more populous nations of the earth.

His philosophy was a simple one: follow your curiosity, seek knowledge and try to leave the world a little better, more peaceful and more worthwhile at the end of each day than it was at the beginning of that day. He also wrote his own music and when he played his pieces on the flute, I was completely mesmerized by the strange melodies and his tonal range. Our relationship together matured as he continued to develop and I did not. He spent more time by himself, often writing or hiking, and even I did not know just what he was up to at any given time. But he was still gentle and kind and it was obvious that he would take great pains to interact with me so that I felt at least equal to him in our discussions. It was a little like a young boy and his

Chapter 6 An Extraordinary person

dad spending an evening together, but in this case the roles were reversed and I, alas, was the boy.

Physically, by the time my son was seventeen he seemed to have reached the limits of his growth, and for this I was exceedingly thankful. He had reached a height of six-feet eleven inches and weighed approximately three-hundred and sixty pounds, distributed such that most of it was in his chest, upper torso, and massive arms, while the rest of his body was rather trim and in proportion with his size relative to his human counterpart.

Actually, considering his ancestry, he was somewhat intermediate in build between the characteristics of the two parent species and his body was not much different from that of an exceptionally tall and well built human weightlifter or wrestler. There was a great difference in potential strength however, in that my son could perform amazing feats of apparent strength with little or no effort. Somehow the great gorilla strength that allowed their kind to tear apart the bamboo forests and plantain to obtain their nourishment has been passed on to my son. It was not so much the amount of muscle mass involved, but the other anatomical and biochemical features he had inherited that governed the way the muscles could be utilized. By the time he was eighteen years old, he was probably the most intelligent and powerful being on this planet. He was my son, and in that year I began to worry about his future, to worry that this enormous polished gemstone of life might

Chapter 6 An Extraordinary person

remain hidden forever. At the same time, I worried as much about the possibility that he would no longer be able to tolerate his isolation and would spring forth, in full bloom on an unsuspecting humanity outside.

I noticed some changes in the way he acted also, and suspected that he had anticipated my thoughts and was also troubled by this matter. I didn't know it at that time, but he was also troubled by the fact that he was the only one of his kind. It was also about this time that I created a new genus in the animal kingdom, one that gave my son an appropriate classification and position in this world. We apes and men all belonged to the suborder *Anthropoidea* or "manlike creatures" of the order Primates. In addition, both of his parents belonged to the same superfamily, *Hominoidea*, of that suborder. His mother was of the ape family *Pongidae*, and I, of course, was of the human family, *Hominidae*, within that superfamily. My son, being decidedly more human than ape, I classified within my own family *Hominidae*. Thus, the problem was to create a new genus in that family which described him properly. I therefore established the genus *Neohomo* or "new man" and of the species *sapiens* or "wise." He was therefore the first representative of *Neohomo sapiens,* or literally, the wise and new man, a title that I think fit him precisely".

Paula and Alan put down the journal to get some more coffee.

Chapter 6 An Extraordinary person

"Could it be", started Paula, "that these beings are responsible for...?" suddenly stopping in mid-thought as if overwhelmed by the magnitude of the revelation.

"Responsible for what?" asked Alan.

"You may know," Paula responded, "that wherever man has become established in the world, he has been confronted by the works of previous beings. These have indeed been magnificent artifacts in many cases, far beyond the capability of the human civilizations which discovered them. We humans have always tried to explain them away as the work of remarkable and precocious human ancestors which had flourished in isolation in various parts of the world. However, there is one common set of characteristics which pervades all of these discoveries. First, all the artifacts would have required superhuman capabilities during the time in which these marvels were constructed. Secondly, there has never been a recorded instance where the remnants of these "superior" races have been found or could be traced back to the beings who actually developed the advanced technologies. These races, in each case, have just seemed to disappear."

"Are you implying that these artifacts are the works of creatures that were not human?" asked Alan.

"Maybe-I'm sure you are familiar with the immense stone statues on Easter Island. These are colossal structures that were hewn out of the tough lava rock and then transported to hillsides at the edge of the sea where they

Chapter 6 An Extraordinary person

were elevated to a standing position to face the sea and eternity. Anthropologists have visited the sites, and have even found statues in various states of completion at the quarries which apparently were suddenly abandoned for some reason. However, it is quite remarkable that these enormous structures could have been shaped at all when you consider that there is absolutely no evidence of an associated tool-making industry. All of this appears to have happened without these beings ever entering the age of metals. Moreover, we have been able to date these construction efforts pretty well and when you consider the hand stone-chipping activity, the weight of these statues and the number of statues completed in the time period, scientists have calculated that the building required the constant effort of more than 100,000 individuals."

"Well, that's not an impossible feat," Alan remarked.

"Perhaps not," responded Paula, "but even today, Easter Island could not support a tenth of that population! Therein lays our dilemma, the human beings required to undertake that enormous building task simply could not have existed on Easter Island."

"But who did live on the island when it was discovered?" asked Alan.

"That's an interesting story in itself", continued Paula.

"It was the Polynesians who had been the most recent migrants and who inhabited the island when it was discovered. However, they apparently had made contact

Chapter 6 An Extraordinary person

with the last few individuals of a previous race called the "long ears", who told them of stories passed down of the "old ones" who had built the statues. Unfortunately, the "old ones", the "long ears" or the Polynesians had the ability to prepare precise written records and so we are left with the vaguest of descriptions passed down by word of mouth and at least in four different languages for many generations. And, as poor a record as it is, this is perhaps the best documented case on record of these formidable beings. The Polynesians also were late comers to the Hawaiian Islands, perhaps settled no earlier than 800 A.D. There they found stone structures and walls of shaped tough lava stone. Their folklore, also passed down for generations, contains stories of the "Menehune" or little folk who had inhabited their islands in the far distant past and who had vanished from the scene.

In South America too, we found marvelous advanced stone houses and other stone works as well as evidence of astronomical techniques that appear to be far more advanced than the capabilities of the people we found associated with them. The immense desert drawings are interesting also, in that they were so large that none who made them could have ever seen them! It was only with the advent of the airplane that they were discovered. Some scientists today believe that those individuals had already discovered flight in the form of heated air balloon airships!

Chapter 6 An Extraordinary person

There are also astronomical mounds in the high desert latitudes of North America which predate the Indians and which we are only now beginning to comprehend. One of the most interesting antiquities however, is at Stonehenge in England. These are ancient astronomical structures which have been found to deal with repetitive celestial phenomena which occur in time in much more than the life span of a single human being. Without benefit of written language then, just who was it that lived long enough to observe, and was wise enough to realize the repetitive nature of these events? Who then had the additional skill and mathematical capability to build these structures? Unfortunately, like the builders of the statues, the walls and the great stone houses, these beings too have vanished from the earth without a trace. "

"Alan, from time to time I can not help but wonder about the relatively long time man has been on earth and the fact that our greatest advances have not been by a slow evolution, but instead by a process in which certain "gifts" have been bestowed by the instantaneous creativity of certain unique individuals or groups. Up to now however, I have always considered these unique and extraordinary individuals to be human beings. Perhaps I have been wrong."

Alan did not respond, but instead looked at Paula for a long moment and then Paula and Alan continued with Dr. Newman's journal throughout that night, wide awake, and

Chapter 6 An Extraordinary person

with no thought of stopping until they knew the entire story.

    Meanwhile, upstairs in the prison hospital, Mangor, with the removal of the intravenous apparatus, had slowly regained consciousness. In pain, blind in one eye, with one lung working and still tethered. In the darkness he closed his eyes and tried to sleep. Tomorrow, he thought, would be soon enough to face the world.

# Chapter 6 An Extraordinary person

Chapter 7
The Encounter

Mary White was born in the same year as my son, and the circumstances of her birth had given her not much more chance for survival in our society than he had. She was born out of wedlock as a result of one of the many chance encounters with which her mother had supported herself. Her father was probably a sailor who took his leave in the port of San Francisco during 1929. Her mother had returned to her own family to have the baby and, due to lack of adequate medical attention at Mary's birth, the infant suffered an unnecessary lack of oxygen for a short period of time just after her arrival into this world. No one at that time noticed the result of this unfortunate event which had produced some brain damage, and as she grew up no one associated her simple-mindedness with those few critical minutes of her first day of life.

Mary's mother, as is so often the case, was devoted to her daughter with a passion that was absent at the time of the events leading to her conception. Mary was a handsome girl and grew straight and tall with sharp features, high cheekbones, a ruddy complexion, and the long black hair characteristic of her mother's heritage. By the time she was sixteen she stood six feet tall, a trait she had probably

Chapter 7 The Encounter

acquired from her father, and was as pleasing to look at as any creature one could imagine. She could not compete in the school system and was placed in classes for the "educable retarded" during her early association with our education process. She was a pleasant soul and would do as she was told, operating at the achievement level of a ten-year old.

While her mother was alive, Mary was reasonably protected from society for they lived in their own isolated artificial world where the demands made on Mary were modest and she could feel a sense of accomplishment. In a sense Mary's existence with her mother was no different than that of me and my son. Unfortunately, Mary's mother died in 1947, leaving Mary then seventeen years old, by herself. There was no one else to look after her and she was thrust upon the world.

It was a not a pleasant story that unfolded the next year. At first she obtained a job as a household helper but after three months, she was attacked by both her employer and his son. Shamed and not knowing what to do, she hid in her room for almost a month, coming out only at night to scavenge food. A resident of the same rooming house, who had known her mother, finally persuaded her to clean up in his small shop in exchange for food. Without her mother, she was no longer able to take care of her own body and fell into a degree of un-cleanliness that was appalling. She did not wash. Her hair became matted and unkempt and her

Chapter 7 The Encounter

dresses and other clothes rotated between her body and drawer. Perhaps this was in her favor in that it kept most people away. Not all however – and she was forced to submit to her landlord each month in order to stay in her room. She longed for friendship but found that women wanted nothing to do with her and the men, those that would talk to her at all, wanted her on her back. She could not bear the loneliness and found herself offering herself more and more often. By the time she was eighteen, she had become a community joke and brutalized by everyone.

Roger Malek, was born in the same year as Mary, was the second son of John Malek and not particularly wanted or planned for at the time. He was a reasonably bright boy and had no particular problems at school or in getting along with others. He grew up in Sacramento where he had many friends and went through the stages of puberty with no more than the usual amount of problems and anxieties. While he did not choose his friends too carefully, he had avoided getting into major scrapes with society and had not ever been in trouble with the police. Roger, like most young adults was a diamond in the rough, abrading society slowly, constantly becoming more smoothed and polished as time went on, but without at any time exceeding the limits that would have evoked societal restraint. Roger's older brother and some of his friends, however, had exceeded these limits and did have records of their deeds –

Chapter 7 The Encounter

or misdeeds – inscribed in societies' book of debts owed and/or paid.

As related to me by my son, It was a warm night in the summer of 1948, the summer during which in the Holy land, man engaged in another war of independence; a night in which my son, greatly troubled, roamed the woodlands under the same stars that shone on the troubles of others. Roger, with his brother and a friend had been to several bars and were in a state of youthful exuberance. Late that evening, they began to drive through town looking for adventure. It was Mary's unfortunate fate to be on the street when the boys pulled over and were able to drag her into their car.

Mary, bewildered and frightened, cried out, but was quickly beaten and subdued as they drove east past the city limits. After taking many back roads, they entered a desolate canyon area and pulled off the dirt road into the night cover of a stand of trees. The preceding portion of the events of that evening I put together from bits and pieces in the published newspaper reports and from conversations I subsequently had with Mary. What follows is a description of what I understand must have taken place from the few things that my son related to me through the years about the evening when chance apparently brought him to the same canyon, not a hundred feet from where the boys' car had stopped.

Chapter 7 The Encounter

The boys knew Mary and at first tried to persuade her to be cooperative and submit to them. She did not know why, but in her simple childlike mind she had drawn a line and this time she refused all of their advances. She was frightened both by what was happening and more so by her own will, which was a strange and new phenomenon to her. She was subsequently pulled out of the car and punched repeatedly as she fought them and tried to run away. Although she was bigger than any of the boys, Mary was no match for the three of them. My son could not leave the brutal scene unfolding in front of him. All of these years he had avoided any other human contact, but he had never faced a situation like this in which he was witness to inhumanity. Roger's brother and friend held Mary up with her arms spread, and Roger, approached from behind and removed her clothes. He then raped her while she pleaded and protested. After he was finished he replaced his brother and his brother replaced his friend in turn, as they all took part in the savagery. Mary lay on the ground after they were done, and stared angrily at the three of them.

Where she got the courage I cannot say, but in clear and steady voice she told them that she would tell the police and they would punished and put in jail. With that, Mary staggered to her feet and began to run toward the forest. Now, I cannot know what went through the minds of Roger's brother and his friend. Perhaps they thought of their police records and feared the consequences of this

Chapter 7 The Encounter

night's rape. Perhaps they were still drunk and not cognizant of what they were doing, or perhaps they were just suddenly ashamed and wanted to eliminate the evidence of that shame. In any event, the two boys lost all sense of reason and began to chase Mary, perhaps not even aware that their intent was to catch and kill her. She did not get far before they pounced upon her and began to beat her with machine-like regular blows of their fists.

My son, watching in the darkness, began to beat his left breast in cadence with the blows, then faster and faster and with both hands, harder and harder until he sprang forth beating his chest, and in a few terrible moments his eighteen years of hiding from the human race were suddenly over. With a single blow across the spine, he killed the friend and then picked up Roger's brother and hurled him against the car with such force that he shattered his skull and most of the other bones in his young body. By that time Roger, who had responded to the fight on an animal level, fearless, not knowing what lay in the darkness of the night, but in selfless defense of his brother came within my son's reach. He grabbed Roger's arm and with all of his power flung him toward the trees tearing muscles, tendons, nerves and skin at the elbow in the process. My son then picked up the unconscious girl and without looking back disappeared into the woods.

The papers the next few days carried the story of the attack. Roger was the only survivor and had with

Chapter 7 The Encounter

superhuman effort managed to drive back using his one good arm that next morning when he regained consciousness. The other arm could not be saved and was amputated at the elbow by the attending surgeon. There was much curiosity about the case and Roger explained as best he could the nature of the attacker. He had seen the beast for an instant before oblivion and knew it was not human. There were many tracks at the scene and plaster casts were made of these as evidence. The nature of the bodily damage inflicted and the unique tracks present at the scene seemed to support Roger's rather implausible story of an animal attack, but further investigations over the following weeks proved fruitless in turning up the perpetrator.

The case also had several curious aspects for which Roger offered no explanation. Roger would not say why the two murdered boys were naked below the waist except for their shoes. He also offered no explanation for the presence of a woman's undergarment at the scene in the midst of the boys' discarded trousers. As to how or why they were there in the first place, Roger said that he did not know anything more as he was in a drunken stupor during the early evening. The fact that the boys were drinking together that night was substantiated by the police. It was also completely understandable that since Mary had disappeared, that Roger should make no mention of the presence of Mary White at the scene of the crime. He had

Chapter 7 The Encounter

nothing to gain, and only more to lose if her involvement in this affair were ever to be discovered. "

Chapter 8
The Wedding

My son had carried Mary back to our house that night and as we attended to her he told me of the events which had led to our involvement. Mary was in very poor condition and during her transport to our house and for several days thereafter she vacillated between delirium and unconsciousness. However, there appeared to be nothing more serious than the bruises around her head and upper body.

I must say that for a moment when my son came home bearing the girl in his arms I had even greater fears. Strange as it seems, the description of the events which finally were unfolded troubled me less than those that I had imagined. We cleaned her up and kept her under sedation through those first few days and my son attended her constantly. The following day the news of the attack was out and although there were excellent descriptions of the scene and the murders and the brutality of those few moments there was no mention of the girl. We reasoned correctly, that as long as she did not turn up, Roger would not mention her either as that piece of testimony could only serve to embarrass him and soil his parents' memory of his brother.

Mary's disappearance became a second and minor matter for police involvement that week, and just like her

Chapter 8 The Wedding

presence, her absence received scant attention. Her room was rented within the week and her worldly possessions redistributed or discarded in less time than that. During the next few days Mary had regained consciousness for brief periods of time, and as I mentioned, being attended by my son she could not help but see him near her bedside. But to my amazement she did not cry out or seem disturbed. For an instant, before she lost all sense of the world, she had seen my son's pounding attack and had felt herself lifted into his powerful and warm animal arms. Sometimes, it is a person like Mary who sees things much more clearly than others, much more simply and directly. My son was her savior and she knew that he would cause her no such harm as the others had done. My son and I decided that for the present she should remain with us. On the fourth day when she was able to comprehend things a little better, we made our suggestion. She agreed immediately and seemed calmed by our offer.

During the following week as she recuperated, we explained our lives as best as we could. I think that she understood, and I was constantly astonished at the way she had no fear whatsoever of my son. In fact she would seek out his company and he would happily be a companion to her on what must have been for him, an almost childish level. She loved music and was especially delighted when he played his flute for her. My son lost all of the sullenness that had become a part of his life that last year. Mary, too,

Chapter 8 The Wedding

proved to be a fine helpmate and took over many of the daily household chores. As time went by it became clear that in our cloistered home and with proper direction, she was a delightful and useful person. The sense of belonging and love seemed to transform her. She had a particular charm and as we learned, a quality of common sense and simple clarity and straightforwardness that seems to be lacking in many so-called normal human beings. I often wondered what she would have had accomplished in life if it had not been for that unfortunate accident at her birth. I must say that Mary's added presence was a tonic for both my son and me.

Roger also had recuperated and shortly after his release from the hospital his family moved away from the Sacramento area. With the passage of each month we felt more and more at ease and began to resume our normal if not unusual style of life. There were no further reports in the papers or on the radio about the events of the attack, or of the strange beast which had caused the deaths. People began to go out at night once again with just the slightest apprehension, although it was several years before overnight outings were planned in the vicinity of that canyon area. It occurred to me that legally, if my son was "human" he would be considered to be a murderer and myself, an accomplice by giving him sanctuary. However, I now had a new respect for him.

Chapter 8 The Wedding

Up to this point he had always responded on an intellectual level, taking time to weigh his responses and his moves so that I never really knew what his true feelings were. This time he had responded on an animal level, just as we all do on rare occasions, and what he had shown were compassion and selflessness. He was a brave and strong warrior who had fought for justice and freedom - even for a human being. I could not help but think first how little we really knew of the society of other animals, of their feelings, their joys, their sense of fairness, and their knowledge of the creation as they considered it in terms we could never understand.

I guess that I should have anticipated what happened next. Perhaps I did, but did not want to think about the natural course of events that might be expected to result from the pairing of two love-starved creatures. In any event, it became quite obvious that Mary and my son were lovers. Mary saw him only as a fine young man who cared for her and treated her kindly. My son, on the other hand, had no possible rules or patterns to follow and merely responded to his own feelings.

As I mentioned, Mary was a big girl, nearly six feet in height and weighed about one hundred and sixty pounds. Although only half as big as my son, his tenderness and care made their love making possible. My second error in judgment was related to Mary's fertility. From her past history, I had assumed that she was infertile since there was

Chapter 8 The Wedding

no evidence that she had ever conceived during the year and a half her life had been promiscuous. In addition, a hybrid, like my son, is rarely fertile and almost never in a back-cross with one of the parent species. However, after Mary had been with us for about six months, it became quite obvious that she was indeed pregnant. No calculations at all were required to know that my son was the father!

I had many long discussions with my son during the following months with regard to the impending birth. We both had absolutely no idea of what its characteristics would be since there was no way to judge which genes would dominate. The child, if viable, would be three quarters human and one quarter gorilla and if the hominid genes predominated it could look exceedingly human. On the other hand, if the anthropoid ape genes were particularly dominant, the child could look very much like its father. We agreed that the child would have to be delivered by us alone even if there were some additional danger to Mary. I was deeply troubled by the possibility that the child might be very much like my son and would also require a special and secluded existence. While it was possible for me to attend to my son for most of his life, at the age of forty five there was little likelihood of my taking care of my grandchild also. Mary, that sweet creature, could not possibly take over these duties.

The baby was born after an easy labor in the spring of 1949 and it was with great relief that I saw that it was

Chapter 8 The Wedding

decidedly human in appearance. The infant, a boy, weighed about nine pounds at birth, with a swarthy complexion and a somewhat brutish look. However, he was no more brutish in appearance than some fully human babies. Fortunately, this child was not a girl. Mangor named the baby Peter.

After the birth we had a new problem to cope with. This baby was illegitimate in the legal sense and would have a difficult time in life unless it had a more conventional position with regard to the human race. It certainly should be registered, documented, and educated so that it could function in the main stream of the society from which its father was excluded by birth, me by choice, and its mother by accident.

There appeared to be only one logical solution. I would marry Mary and raise my grandson as my own child. After several weeks of confinement, Mary and I took the baby to Reno, Nevada where we were married in a simple civil ceremony, and at that time also recorded the previous birth of Peter M. Newman as the legal son of Joseph and Mary Newman. When we came home my son slept with my legal wife that evening, and for many peaceful evenings to come."

Chapter 9
Family

We all enjoyed the baby immensely and for the first time in my life I really felt happy with my lot. The world outside was not in such a happy state however, with the start of the Korean conflict and the deaths of thousands of men and women of several of the human races. The next two years, I must say, were pleasant and quite surprising and enriched our lives considerably. Matthew M. Newman was born in 1950 and Andrew M. Newman was born in the following year. With three babies born within a period of three years, there was no end of activity day and night. You may wonder why each of the children had the middle initial "M". Actually, it was my idea that the children should carry some identification of their true father in their given names and so each was given the middle initial "M" to represent "Mangor," the name I had called my son.

Like Peter, Matthew and Andrew were decidedly human in appearance. We had taken the chance that this would be the case and had placed Mary in a clinic when she had reached full term - a period that incidentally was only seven and a half months of gestation. Both Matthew and Andrew were covered from head to foot with sparse black hair at birth, a condition not uncommon in human children

Chapter 9 Family

but usually not as pronounced. As in their fully human counterparts, this vestigial cover was gone within a few days. In no other way that I could discern was there any other mark that could differentiate them from other babies of the human races.

My son loved and played with them, cared for and fondled them incessantly, and they came to love him. These children were quite different from their father however, in that they were decidedly normal, exhibiting none of the precocity that had characterized my son's development. In fact, it soon became clear that these children would grow up to be quite average in their level of achievement. This solved one problem, that of their attendance at school, but, presented us with another. It was inescapable that, as they interacted with human kind, they were bound to compare their homes and fathers with others. Should they be told when they could comprehend the facts, just who their real father was? How would the presence of a talking and quite unusual animal-like being in their home, roaming freely and carrying on everyday activities associated with their human counterparts be explained? We felt that we had at least six years to cope with this problem before sending Peter off to school and decided to chance the truth! Our home was accepted by the children as warm and friendly to their young eyes and minds and our initial apprehensions proved to be unwarranted.

Chapter 9 Family

We counted on the years and "reason" to prepare the boys for a secret which they would have to carry for the rest of their lives. I am constantly amazed at the adaptability of the human species and fortunately our hopes were fully realized. The children accepted their true parents for what they were and as each entered into human relationships they did not reveal the other side of their lives. Fortunately also, although the area had seen some development, our home was still quite isolated, a fact that allowed us to carry on our lives with only some inconvenience. The boys had all of their other human contacts away from home, and Mary and I carefully avoided any other social contacts. I represented the boys' family at any function which could not be avoided, and continued to ply my trade in distant farm communities.

Throughout this period which saw another senseless war in the middle east in 1956, my son continued to work on his journals and spent a good deal of his time each day in thinking and writing. The journals, although incomplete, are rich in their analysis and each subject that he covers is projected beyond man's present knowledge. Perhaps the fact that he is alien to our world gives him a much clearer outlook and perspective. I tried to convince him to submit some of these journals for publication through me, as I had done for his paintings, but he felt that it would serve no purpose for the present and he would, if anything, rather leave them for posterity. He is probably quite right and I

Chapter 9 Family

hope that mankind will benefit from my son's rich and inspired legacy in the future and for all time. In the year 1966 our lives were shaken when Mary, just past her thirty-sixth birthday, suddenly died. This gentle and loving creature took sick one day with stomach cramps and nausea, but not thinking it serious, we treated her for a common intestinal virus. Very quickly though, she ran a high fever and by the time I rushed her to the hospital she had developed pneumonia with a fever that reached 105 and that would not respond to treatment. She died that night, the cause of her ailment unknown.

The funeral was very simple, attended only by myself and her sons, and she was buried in a small cemetery several miles from our house. In the days and months that followed, my son was a frequent visitor to her grave. At first my son seemed to take Mary's death better than any of us but as time went on I began to see changes. He became increasingly morose and began to spend much more time outdoors, and there were often days and nights when he never returned home. Although I never thought that he would get into any trouble from which he could not extricate himself, Without Mary however, my life was very lonely when he was away. The boys were in their teens, with Peter already seventeen years old, and their involvement in human activities in the outside world grew more intense. They seemed to have little time to spend with an old man, for at sixty-three years of age I must have

Chapter 9 Family

seemed like a relic in their world. It was only some time later, when my son brought all of us together and presented his plan, that I learned where my son had been spending all these hours and days away from our home.

At first, roaming the wilderness areas of the high Sierras, moving from ridge to ridge and wandering through the high Alpine meadows, he had only wanted to be by himself away from home where he could perhaps forget his pain at losing Mary. But after a while, he conceived a notion that would solve many of the problems we had discussed earlier in our lives. By this time, it had become obvious to Mangor that he would outlive me. Without Mary, and wishing to remove the burden of his existence from his own sons, he had decided on a plan that would cause none of us anxiety, and would provide him with his needs for the rest of his life.

The plan, as presented to us was simple enough. He had found a suitable cave high up in the mountains where the isolation of the mountains would afford him peace and quiet, if limited comfort, in which to live out his life. We would spend the next few years making it ready and after my death, he would retire to that hermitage severing from that time on all ties with humanity. Nor would he ever see his sons again, a choice he made for their good - and perhaps for his own. Often, as I reflected on this decision, I came to the understanding that it was the best solution to a very knotty problem, a solution indeed that only my son

Chapter 9 Family

could have made for neither I nor his sons would ever have thought of it, or had we done so, brought ourselves to recommend it.

The move was carefully planned, for at the outset my son realized that if he were to live alone and isolated for any amount of time and under such very primitive conditions, he would have to be self-sustaining. It would be necessary for him to obtain and store his food during the summers against his winter's needs. Because there would be no possibility of medical aid, if he were to be hurt or otherwise incapacitated it would probably mean his end. He would have no friends or family to count upon and no one to talk to, but hopefully, his isolation would also mean that he would have no enemies.

My son decided that he would take only some tools and a medical kit, and that he would make all other things that were necessary - or make do without. His knowledge of biology and his immense strength, I was sure, would indeed enable him to survive under the harshest of conditions. While these were the only things that he took to assure his physical survival, he did on the other hand require much to keep him occupied in his solitary existence. This did not present any great problem because he had after all been alone for most of his life, and so he naturally elected to bring things that he had loved when he was by himself, his textbooks, journals, and writing tools, his flute and several pairs of spectacles. His only other comfort would be several

Chapter 9 Family

lamps and enough oil and wicks to keep them going for many years.

Each summer when the snow finally melted for a short time in August, we went to his cave on a ridge above a high meadow and prepared it for the future. We built some crude but strong furniture and a door to keep out the elements. The cave had been carefully chosen and was quite amenable to being made into a snug and functional retreat. Now, at the time of this writing, Peter is twenty, Matthew, nineteen, and Andrew, eighteen. Early this year my grandchildren enlisted in the army and have been sent to Indochina for some sort of International police action. As before, and as I have for most of my life, I live quietly with my son.

Joseph Newman, May, 1969"

Chapter 9 Family

Chapter 10
Diaspora

The sun was just rising on the new day and Paula and Alan, having completed Dr. Newman's journal, sat back for a long while enjoying the quiet and solitude of that special hour. Both reflected on what they had learned during that evening's reading, and neither knew quite how they would deal with it. Paula felt very close to the creature at that moment and sensed an impending tragedy. Alan's thoughts carried him back to the creature that lay upstairs in pain and helpless. At that moment they alone other than his three sons knew that the creature that had been hunted and captured was indeed a sensitive and rational being.

"Well," said Alan after awhile, "what do you think?"

Paula looked at him for a few moments before responding.

"I don't know, but I guess that Dr. Newman died sometime in 1969 and his son, alone for the first time in its life, went into the mountains according to his plan."

"That makes sense," said Alan. "It accounts for the increase in strange sightings in this area during the past fifteen years and for what ultimately led Roger to that high meadow."

Chapter 10 Diaspora

"He must have planned well to have survived the winters up there for so many years" said Paula." And the handwritten journals, the ones that were signed 'M. Newman, and that we thought were those of Dr. Newman, those were really his son's work, his major occupation during his long exile away from other rational beings."

"It's six o'clock" said Alan. "I'll order some breakfast and, while you get yourself tidied up, there is something that I have to take care of upstairs."

With that Alan left and made his way up to the prison infirmary.

Mangor had been awakened by the sun shining on him through the cell window. For a moment he had forgotten where he was but the pain and stiffness of his injuries made worse by his bound condition quickly brought him back to reality. Despite his discomfort, he found it pleasing in the quiet of the morning to be awake and breathing in the cool fresh air. He felt curiously detached from his strange surroundings and for a moment almost oblivious of his physical state. After living in solitude for so many years and although he himself had a family, Mangor still found it very difficult to relate to other individuals.

He also remembered that except for the day he was shot, it had been years since he had even articulated words. At first during the early part of his isolation in the cave he had found himself talking to other animals and even to inanimate objects. But as the years went by, the words he

Chapter 10 Diaspora

used remained only in his thoughts and he stopped using the spoken language altogether. He wanted to talk to his captors but found that the formation of words was now slow and the additional effect of the recent surgery made attempts at articulation a painful process. Just then he heard some activity outside of his cell, and shortly Alan quietly unlocked his cell door.

"Mangor" he said, "I know who you are. Please don't be afraid of me."

With that he proceeded to unlock Mangor's hands and feet. Mangor made no reply and made no attempt to rise. Alan, still not quite certain about what he was doing or even about the truth of what he had read, slowly backed away and left the cell locking it behind him.

"There, that should make you a bit more comfortable", he said, "I'll see what I can do about improving your living quarters. See you later."

With that Alan returned to the outer security zone and spoke to the deputy on guard.

"I've unlocked the creature's chains and he may be up and about soon. Don't worry about anything – I'm sure you're perfectly safe. Now, there's a cell adjoining the one that our captive is in. I want you to put a regular bed in it, along with a table and the biggest strongest chair you can find. Also, see to it that the bars on the aisle are completely boarded up, including the door. I want that cell to be

Chapter 10 Diaspora

completely private as soon as possible. Are there any questions?"

The deputy shook his head as a somewhat confused expression spread across his face. Then abruptly he shrugged.

"When it's finished, Alan went on, "give me a ring downstairs in the exhibits office. There's one more thing. Before you call me, order two big breakfasts and have them brought up here and placed in the adjoining cell."

Before the deputy could voice the questions that sprang to his lips, Alan turned and went back downstairs to Paula and breakfast. They sat down to eat with great relish, and the coffee along with rolls and eggs provided seemed particularly good that morning.

"I've been up to see him," said Alan. "He seems to be okay and I - now that I think of it, maybe stupidly - unlocked his chains."

"Did he say anything?"

"No. I talked to him and told him that I knew who he was, but I didn't give him much of a chance to answer. At any rate, I'm having the adjoining cell prepared so that he can have some seclusion. It should be ready in a few hours and we'll see if he responds then. He didn't cringe from me, or try to attack me either. He only looked at me."

"Well," said Paula, "where do we go from here?"

Alan thought for a moment and then replied.

Chapter 10 Diaspora

"First, there's some paperwork that needs adjusting. Then, I'm going to try to verify the story and see if I can get some documentation on Dr. Newman and his wife and their "sons". I'll try to locate the boys - they must be grown men now - but if the story is true, it may be extremely painful to them and they may out-and-out deny any relationship with Mangor. I'm sure that I would in their position. Unfortunately, I don't see how this whole affair can be carried out in secret. Something this extraordinary is sure to get out before long. By tonight I also want us to decide just how we are going to present the apparent facts in this case. I'll keep others away from the evidence, but it's not right to keep it to ourselves for very long."

Paula was silent for a while.

"What do you think will become of him?" She asked. "Not in the sense of anthropology or biology, but as Sheriff, what can you tell me about the legal ramifications in this matter?If Dr. Newman's journal is true, then Mangor is an extremely sensitive and rational being and he should be treated as such, not as a wild brute to be kept caged as he is now. I know that we scientists have our work cut out for us, but how do we get him to walk out of that cell upstairs?"

"You have me there," replied Alan. "As I see it, there's no way that I can just let him go. In the first place, if we consider him a wounded animal, he comes under a host of ordinances that control the way he must be handled. In any

Chapter 10 Diaspora

event, even if he is who - or what we think he is, he would have no rights and couldn't be let loose to roam the streets at will. As a matter of fact, at this point, the District Attorney can order me to deliver him to the zoo, have him dissected, or even have him stuffed as a type specimen. They may even have me destroy him as a known killer animal if they believe the allegations made in the Malek case."

Alan paused for a moment, considering and mulling over the facts as he knew them.

"But of course" said Alan, "we may be quite premature in even talking about what we have upstairs. It may be, although it would surprise me, that Dr. Newman was just plain crazy and that the beast upstairs is just what he appears to be. Or, Dr. Newman might have been just half-crazy and it can really reason and talk and is what he said it is. But, perhaps it is not a civilized genius but instead a clever lunatic which can kill for no apparent reason. And maybe I'm all wrong and he really is as advertised. How can we judge which is correct at this point?"

"On the other hand" Alan mused. "If your assumption is correct and if by some quirk in our system of justice, he is considered a "person", then theoretically - no, actually - he would have to be considered a captured criminal and tried for his crimes. With evidence of the encounter in Dr. Newman's journal and the physical evidence from the scene of the crime, my guess is that he would be convicted."

Chapter 10 Diaspora

"Certainly you don't mean that he could be tried on those charges," cried Paula. "That happened thirty-six years ago. What about the statute of limitations? Also, would Roger want to press charges and jeopardize himself?"

"Sorry Paula, as far as Roger is concerned, he's in the clear on two counts. First, there's no plaintiff in the rape case, and second, the statute of limitations has run out on that offense. However, in a murder case there is no statute of limitations. In addition, remember that even if Roger does not press charges, the State must pursue this case since it is still officially open and we have found new evidence."

"My God," exclaimed Paula.

"Look" said Alan, "there's no good reason for us to get so excited at this point. Who knows where this is all going to lead. And, even if your suspicions are true, it's not as bad as all that. Since he is technically not a person, the State can't institute any proceedings against him under our system of justice. If nothing more is done, he'll probably be treated as a rare animal which means that without any rights, he'll be used in any way seen fit by his captors. What it all means is that he'll probably be confined for the rest of his life and tested and probed as a specimen in every imaginable way. However, knowing what I do now in the light of Dr. Newman's journal, I'm not sure that he will survive such confinement."

Chapter 10 Diaspora

Paula was perplexed and said nothing for a long while. Outside the sun was rising and the day's noisy activities had begun.

"What if I force the issue?" asked Paula, "What if I, on behalf of the citizens of California, sue the State for his release?"

"Well, in that case responded Alan, you'll probably precipitate a hearing to deal with the basis for your petition and the entire question will have to be aired in a public forum. Your suit may be dismissed outright as irrelevant, or it may result in a careful examination of the issues involved. I'm not sure which way it would go. Anyway, I think that it's too early to make any plans.

"Would you like to go up and see our guest now?" He's not uttered a word to anyone yet."

"Sure, let's go," replied Paula. "There are a few things I want to be sure of before going off half-cocked."

Upstairs, the carpenters had finished the work that Alan had ordered. The adjoining room was now sheathed in plywood and had been properly furnished. Alan took the two breakfast trays inside and placed them on the table. In the adjoining cell, the creature had assumed a sitting position and calmly looked out of his cell at Paula and Alan.

"My name is Paula," she began somewhat hesitantly "we know who you are."

There was no response.

Chapter 10 Diaspora

"Look, we mean you no harm," said Alan, "I've prepared the next cell for you so that you can have some privacy. I'm going to unlock the door between the cells now, and you can move in when you like. There's some breakfast for you."

Still there was no response although the creature seemed to nod. With that Alan entered the next cell, unlocked the adjoining door, and then retreated into the corridor.

"I think that we had better leave him alone for awhile" Alan remarked to Paula. Anyway, I have lots of things to attend to and I suppose you do to - including getting a little rest. How about meeting me back here at about five? We're going to have to decide about this thing tonight, one way or another because I'll have to make my report tomorrow morning."

Paula, feeling the effects of the long nights work, agreed and left to sort herself out. During that day, Alan gave orders to his staff to try to locate any records they could find on Dr. Newman and his wife, as well as to try to locate the whereabouts of his three "sons." He also placed a call to Roger and, not to his surprise, found that Roger was quite cool when he heard that the creature was going to survive. Alan felt it was not the right time to bring up the matter of their last encounter and in fact began to have fears about the creature's own safety.

Chapter 10 Diaspora

Meanwhile, on the way back to her room Paula contacted her lawyer and requested that he look into the preparation of a motion to release a prisoner being held without charges. She then took a hot shower and slept until late afternoon. They met again at five, as planned.

"I think that I've located Mangor's sons, announced Alan although no contact has been made as yet. I've also been able to document the births of both Joseph Newman and Mary White out of the San Francisco office and we've sent off to Reno for documentation of the wedding between those two. We also came up with a record of a death certificate issued for a Mary Newman and the date appears to be about right."

"Has there been anything new upstairs?" inquired Paula.

"Yes, he apparently took up residence in the adjoining room shortly after we left, and when I went up there about one o'clock I found the empty trays neatly stacked at the food service door. I saw him lying on the bed and spoke to him, but he didn't respond to anything I said. Do you have any thoughts about tomorrow?"

"Yes, it's pretty clear that we have no choice or any right to withhold information. However, I suggest that we brief the Governor first and let him decide on how to go about it. He may want to contact Washington before taking any action. Also, I strongly suggest that we warn the Governor not to mention anything about the possibility of

Chapter 10 Diaspora

'Mangor' having living children or about the relationship between Roger Malek and Mangor at this time – a great many people's lives might be disrupted by a disclosure like that."

"What have you decided about what you are going to do about his release?" asked Alan.

"I'm going to take legal action if necessary." replied Paula, "and I'd also like to see him once more tonight to try to communicate and to let him know what's in store," said Paula.

"Just a moment", Alan said, "before we go up, there's something I'd like to get from the evidence room."

They went downstairs and Alan got the silver flute and then the two of them went upstairs. The supper trays had just been cleared and Mangor was standing near the window in the open cell. Paula proceeded to explain what course of action she was going to take, including the disclosures that came to light about him and her possible legal action to free him. Mangor seemed attentive but still made no reply or comment on Paula's proposal. After Paula was finished, he turned and looked out of the window once again.

"Mangor," began Paula, "please tell us what you want done. Unless we're sure of the truth of the things that Dr. Newman wrote about in his journal, we simply can't take any action at all."

Paula said to Alan, "I'll be back in a minute."

Chapter 10 Diaspora

With that Paula went to the outer security desk and returned with the flute in her hands. She then extended her arm and the flute through the bars and into the cell. Mangor turned and looked at Paula once more. After a moment he started toward Paula with labored steps, and, as Alan placed his hand on his gun, Mangor gently took the flute from Paula's outstretched hand.

"Thank you," Manger said haltingly and with that he placed the flute to his lips and began to play.

"Thank you," exclaimed Paula as she smiled and turned to Alan.

Paula and Alan then listened to Mangor, outwardly a brute, as he played extraordinarily delicate and wistful melodies on his flute for quite a while. Then, smiling at one another, they quietly left.

Chapter 11
Revelation

The next few days were frantic, filled with questions and answers, decisions and indecisions. Alan had notified the Commissioner of Police that he had in his custody an individual whose potential status was, at the very least, debatable. The Commissioner, in turn, after the first shock of the Sheriff's revelation had worn off, got in touch with the State Attorney General who made an appointment with the Governor. A meeting was quickly called in which the nature and possible ramifications of the situation were to be explored.

"Sheriff," began the Governor, "I've heard stories about this individual in custody from several second-hand sources, and I would like you, as the person most intimately involved, to present the plain facts in the matter. However, if there's any truth in what I've already heard, I can tell you now that I have no idea about what to do about our, how shall I put it, guest."

"Well," began Alan, "the individual had been shot and was close to death when we were notified and entered the case. Our first consideration was to try to save what appeared to be a rather unique and strange looking animal and of course you know that with the terrific cooperation

Chapter 11 Revelation

we obtained from the human medical trauma team, the creature was indeed kept alive. It was not until some time the next day that we read about the alleged origin and characteristics of the individual in question and knew that this was truly a remarkable being.

Since that revelation, I can now tell you that Dr. Paula Chapman and I have made contact with the creature and have verified some of Dr. Newman's statements. The creature now in our jail speaks English, plays a musical instrument and appears to be a civilized and rational being."

"Sheriff," responded the Governor, "that is basically what I've heard before and even now coming from an eyewitness, I find it hard to believe. I want to meet with this individual myself as soon as our meeting is over. I would also like to review Dr. Newman's journal myself after I return from visiting the jail. Will you also see to it that several copies of the original document are made before it is removed from the evidence file."

After some additional discussion about security and the creature's accommodations, the Governor then proceeded to order the State Department of Justice to give him a report on available briefs dealing with possible precedent in a matter of this kind and with these initial actions taken, he and the sheriff left for the jail to meet with the 'creature'.

It took several days to prepare the report, during which time unsubstantiated rumors began to spread through every

Chapter 11 Revelation

department of government. Even the press began to issue statements with regard to certain strange happenings. The results of the legal search had failed to turn up anything specific, but the lawyers did suggest making an inquiry to Washington. Feeling that he had no firm position of his own and taking the course of least resistance, the Governor contacted the Federal Department of Justice in Washington for guidance. The response that arrived three days later indicated that the government would send several observers, but that the primary jurisdiction in this case was with the State. On the fifth day following the sheriff's disclosure, the Governor called a conference to which leading members of the State government, scientists and the press were invited.

"Gentlemen, I've called this meeting today to present the facts in what has become an extraordinary case. You have by now probably all heard about the strange ape-like creature which was captured several weeks ago. Undoubtedly most of you have also heard the rumor that this creature is a rational being and has the power of speech. I can tell you now that this is all true. In the past week I myself have had the opportunity to meet with him on several occasions and have had lengthy conversations with him on the validity of a certain document, a history so to speak of his life.

At this point, gentlemen we have no reason to doubt any portion of the document. What it alleges is that the

Chapter 11 Revelation

individual in custody is the result of a successful cross between a female mountain gorilla and a human male via the technique of artificial insemination in which, as you may know, sperm is introduced mechanically using a surgical device. The cross took place in 1929 and the human donor in this case was Dr. Joseph Newman, the document's author. Therefore, the creature is the hybrid son of Dr. Newman. His given name is Mangor; he is fifty-four years of age and was born in and has lived in the State of California for his entire life. The details of his remarkable life will be released to the public only after certain legal matters are cleared up. Well, that's all that I have to say for the moment, but I will answer some questions."

As if suddenly released, the reporters clamored for his attention. Hesitating only a moment, the Governor nodded to acknowledge one of them.

"Sir," the chosen reporter began "you say that you have had conversations with the animal. Were these face to face? Is he not restrained?"

"Yes, replied the Governor," I had face to face meetings with Mangor – Mr. Mangor Newman – and I can assure you that he is quite civilized and will not cause any harm. For legal reasons, however, he is confined to a suite in the prison infirmary."

"Governor," can you tell us how he came to be shot?"

Chapter 11 Revelation

"Well, at this point, all I can say is that he was mistaken – and understandably so – for a wild animal as he was walking in the mountains."

"Sir; this is certainly an amazing story and, if true, will represent an important advance in the study of anthropology and also will certainly pose some interesting questions in other areas of science. My question is this; what will be done with the creature? Who will get to study him first?"

"I'm afraid," replied the Governor, "that the question of his future has yet to be resolved. I mentioned that Mangor is a rational being. From my own observations I can tell you that he is more than that, much more! He is an accomplished musician and a scholar of exceptional ability. He reads and speaks eight languages, including Russian and Chinese. We have had experts in various fields examine his unpublished writings. They all agree that he has great genius and that his insights in these fields will be of great value to the human race. In short, I'm not sure that we are going to study him at all. We may in fact be too busy learning from him."

More hands shot up, but the Governor had given out all the information he had planned to, and he shook his head.

"That will be all the questions for today. My office will issue several bulletins within the next week or so to fill in some of the missing details. Thank you for attending and for your patience."

Chapter 11 Revelation

By the next morning, the news of Mangor's existence and his unusual parentage had reached around the entire world. Within several days there were official offers of asylum from five countries including Sweden and Russia. Uganda conferred citizenship on Mangor and demanded his extradition, which was as a matter of course denied.

Several major religious orders and lay organizations sent letters and envoys to both Washington and California demanding the creature's elimination in order to remove a grave offense to God. Letters and editorials to newspapers were running pro and con with regard to the truth of Dr. Newman's story in about equal numbers and special interest groups were petitioning both the President of the United States and the State Governor.

The Sierra Club wanted him returned to the mountains; the National Academy of Health and the National Science Foundation wanted to issue "requests for proposals" for several specific studies in medicine and biology to be performed on Mangor in the following year. A religious order in southern California even pronounced him their savior and wanted him released immediately to take over their church. All during the following week the Governor was besieged by the faithful, the informed, the misinformed, do-gooders, no-good doers and just plain nuts of every description.

There were two serious matters that had still not been disclosed. First, the matter of Mangor's sons, and second

Chapter 11 Revelation

the murder case still in the police active files. In addition, the Federal Department of Justice had entered the matter and had advised the State Judicial department of some legal problems that might surface in this case. However, it was made quite clear that the matter was to be left in the hands of the state courts.

Late in the afternoon on Friday of that week, Alan was able to arrange some time and went to visit Andrew Newman at his home. The house was extraordinary, especially for the area, being made of large stones and with a heavy slate roof. It was a reflection of the fact that Andrew had started a rather successful restaurant business since his return from Vietnam. He had lived on the same site for a long time before, and when the money was available, he elected to tear down the old house and build the new and better one in its place. The sheriff knocked.

"Who's there?" asked Andrew.

"It's Sheriff Garfield," he responded, and with that Andrew opened the door.

"Hi, I've been expecting you" said Andrew, "but I'm afraid that our meeting will have to be short right now since I'm due at the restaurant."

"I'm sorry that I was delayed," said Alan, "but it shouldn't take too long to discuss the matter I mentioned to you over the phone. As you probably know, we have in custody a rather unique individual and also a journal purported to be written by a Dr. Joseph Newman which

Chapter 11 Revelation

reports on certain aspects of your family that we find hard to accept. Do you know anything about the journal in question?"

Andrew poured himself a drink and then responded.

"No, I have never seen the journal to which you have referred, nor have I ever been aware of its existence."

"I see," said Alan, "I will of course let you see the journal any time, but for the moment, let me state that the journal indicates that the individual that we have in custody lived with your family for many years, and more importantly – and this may be quite a shock to you – it indicates that the individual in custody, and not Dr. Joseph Newman, is your father."

Andrew, sitting, took a deep breath and was quiet for several moments. Then he responded.

"Sheriff, I have never been aware of the journal which you have described, and as far as I am concerned I am the son of Dr. Joseph Newman and no other. The writings of my father and the results of his many experiments are not of my concern. I'm afraid that it will be up to the scholars to interpret them and decide on what is right or wrong. All I can tell you is that my father was a very gifted man and no course is open to me other than to abide by his decisions and to live by his rules. If, on the other hand, what you have alleged with regard to the individual in custody is true, then what earthly purpose would be served as far as my family is concerned by my verifying that strange story.

Chapter 11 Revelation

Surely, you wouldn't have expected me to jeopardize all that I have worked for just for the sake of the truth? That is, if it were the truth."

"Andrew," the sheriff answered "I heard what I expected to hear, and perhaps understand even more than you think."

"Now, if you will excuse me," said Andrew, "I must get to my business."

The sheriff nodded and put on his hat.

"Thank you for seeing me, Mr. Newman, we will be in touch with you. Goodbye for now."

"Goodbye sheriff," said Andrew as he closed the door behind his visitor.

Alan then returned to his office to report on the meeting. Later that evening, he had dinner with Paula and they discussed the problem that confronted Andrew if the journal statements were in fact true.

After several days of soul-searching the Governor called a second conference to announce the States' position and the Attorney General's decision with regard to Mangor. He had also contacted Dr. Newman's sons, none of whom would acknowledge being a descendent of Mangor. All three had families of their own and their reluctance was understandable. While they did not admit to any relationship, interestingly none of them denied it either.

A still unresolved problem was the matter of the murders and the possible implication of Mangor. The

Chapter 11 Revelation

sheriff had indicated that he would take no action unless Roger Malek requested it, and to date Roger had given no indication that he had any more interest in the case. Taking these matters into consideration and also considering the nature and degree of public interest and involvement in the case, the Governor had visited Mangor and had explained his position. Mangor did not like what he heard, but offered no alternative. In fact he seemed to be a bit disinterested in the entire affair. The second news conference was well attended but short and no questions were permitted.

"Gentlemen, there have been many developments in the case of the so called 'creature' in the past two weeks. I have called this meeting in order to issue a statement with regard to the position that is being taken by the State of California. Considering the many aspects of this matter and the best interests of all involved parties, we have decided to confine Mangor to an institution for the rest of his natural life. His confinement will be under what I believe to be humane conditions, where he will be allowed to pursue his interests and have limited interaction with other individuals. Hopefully, in several years, and with a minimum of impact on Mangor, we will have learned what we need to know about the results of this bizarre experiment, and at that time not have to subject Mangor to any more unnecessary suffering. A full statement will be issued in due course. Thank you."

Chapter 11 Revelation

Paula, sitting in the rear of the auditorium stood up and went directly to her lawyer's office. There was no question in her lawyer Mr. Darrow's mind about this case. The next day, Paula initiated her suit against the State of California for the immediate release of Mangor, a "person" being held in the State Prison without due process of law.

Chapter 11 Revelation

Chapter 12
The Gathering

Mangor heard the news of the Governor's statement and was dejected. It was not that the course of action taken was not reasonable or was unfair, but he had never felt so alone. Not even when it was only he and his father living isolated from the world around them. The worst of Dr. Newman's fears had come to pass and there appeared to be no place in society where his son could function during his brief time on earth. The worst part of it all was that he no longer seemed to care. Whereas he had always been active in one pursuit or another, even during his self-imposed isolation high in the mountains, he now found himself doing nothing.

He had just given up. During a good part of each day he slept in his two cell apartment. When awake, he would find himself looking blankly at a wall or staring out of the cell window into space. Mangor's feelings were not unique and were shared by others of the human species. It was a trait exhibited by many who, for one reason or another, felt that they could no longer cope with the frustrations of the situations in which they found themselves and regardless of the seriousness or triviality of those situations as seen by others. At this point Mangor resigned himself to the idea of

Chapter 12 The GATHERING

spending the rest of his days under controlled conditions in an institution of some kind and was even looking forward to the time when some sense of order and relative isolation would return to his life. He understood, but was at first disappointed in his sons.

Hearing that the hunter was one of the boys who had brutalized his mate so many years ago also brought back things that he would rather have not remembered. He had long ago decided not to hate those creatures since, for if it were not for their actions that evening, he might have missed the best part of his life, the gift of Mary.

Even now, he could not hate Roger for what had happened. Mangor had murdered and maimed in a fit of anger, and in the world of his ancestors retribution was simple, direct and correct. He did not want to face Roger again or think anymore about that night. Asylum also seemed to be more attractive than having to cope with the world at large. It was not that he hadn't thought about it. He had always had a fantasy of being accepted by humanity and being respected and allowed to collaborate with the scholars whose works he knew so well. However, the events of the past weeks after his secret was out, showed him another side of humanity in which many called for his execution and which exposed his fantasy for just what it was.

A final option had of course come across his mind, that of going back up into the high mountains. Even in his

Chapter 12 The Gathering

weakened condition and without the sight of one eye, he felt that he could escape. But the thought of solitude again, of not even the companionship of his books, and of the never ending hunt for him, dissuaded him from that course of action. And so he retreated from reality and for a moment left his fate in the hands of others.

It was several days later that Mangor received news which was to cheer him. His sons visited him and brought their wives. Mangor's thirteen grandchildren ranged in age from six months to twelve years of age, each one-eighth gorilla but none any worse for it. Each of Mangor's sons had made his decision to tell his wife about their heritage independently. As Mangor had accepted his heritage without question and Mangor's sons theirs, their wives also accepted their situation. It is an interesting fact that people who do not feel inferior cannot be made to feel so by any negative account of their heritage. On the other hand the positive aspects of ones heritage are quickly added to their own worth and people burst their buttons. So it was with Mangor's family, they did not see the gorilla in their constitution but only saw the genius of Mangor. Mangor was no longer alone. There were now seventeen of his constitution walking the earth and his sons assured him that they and their families would bear testimony to that fact. Mangor's eyes brightened that day and he played his flute for his sons once again.

Chapter 12 The Gathering

The second bit of good news for Mangor came from Paula and her attorney. Paula explained the reason and basis for the suit she had filed on behalf of the people of California. Mangor had grown fond of Paula and understood her reasons. Had it been a week before however, Mangor would have been displeased with the idea of her taking this independent action. But now, he felt right about it. He would fight for his place in the world on any and every battle field and he had decided that he would not die behind cell walls just because he was different.

The attorney had explained some of the technical problems that would have to be faced and just what the suit would bring about. Since the basis of the suit was that a person was being held without due process, there would have to be a ruling on whether Mangor was indeed a person or not. Since there never had been a case of this sort, he anticipated that the state prosecutor and the defense attorney would be on equal grounds and he hoped that the arguments in favor would win out.

It was also pointed out by the attorney that winning this decision would not allow Mangor to go free immediately since the very winning of that point immediately subjected him to being charged in the courts as the perpetrator of the double murder those many years ago. However, both Paula and the attorney felt that the extenuating circumstances in the case and the lack of an eyewitness would cause the charges to be dropped. Mr. Darrow also informed Paula

Chapter 12 The Gathering

that Roger Malek had been approached by the state prosecutor and had refused to press charges in this matter.

Mangor, buoyed up by his new friends and family, and by the many unknown supporters he read about in the nation's newspapers and in the letters which came to him, began to write poetry. The next three months leading up to the hearings were filled with activity and companionship for Mangor; more than he had ever known in his life.

Chapter 12 The Gathering

Chapter 13
Trial

THE FIRST DAY OF THE TRIAL

The hearings on Dr. Paula Chapman's case against the State of California began on January 15, 1985. Present in the crowded courtroom were Dr. Chapman, Sheriff Garfield, Mangor Newman and his three sons Peter, Matthew and Andrew.

The judge assigned to the case was his Honor, Henry Dawson. In this matter before the court, Dr. Chapman was represented by her attorney, Mr. Raymond Darrow Esq., and the State was represented by Mr. John Bryan Esq.

Members of the press and other interested parties to the proceedings filled the courtroom and the upper public gallery completely and there was a substantial overflow into the corridors of the courthouse. The proceedings were called to order sharply at nine o'clock in the morning with all due formality.

"Ladies and gentlemen," began Judge Dawson, "we are gathered here today to rule on a suit that specifically charges the State of California with detaining a person in a penal institution without due process, thereby depriving that person of their God given and legal rights.

Chapter 13 Trial

The peculiar nature of this case, specifically relating to the "person" involved, has required that we have an unusual type of hearing. At the outset, since there may be an outstanding warrant on murder charges against the individual involved in this case, a Jury of his peers should be in order. However, since the special circumstances of this case make that impossible, the State has elected to have an adjudicatory hearing instead. In this process, each side will present the evidence in their respective cases and the Judge, after consultation with members of the State Department of Justice, will issue a decision in this matter.

Each side can present exhibits and or witnesses with regard to pertinent aspects of this case and I will try to be as flexible as I can in extending the scope of testimony offered in the interest of best serving justice.

All aspects of a proper trial including initial examination cross examination and witness recall will be maintained. In addition, I expect this hearing to meet appropriate behavioral standards on the part of everyone involved so that the issues can be properly addressed as objectively as possible. As background in this case, let me state the position that has already been taken by the State of California.

The "being" at issue is an individual by the name of Mangor Newman. Mr. Newman, because of his special attributes, has been made a ward of the State for his own protection and well being and in order to avoid certain

Chapter 13 Trial

inevitable disorders which would occur if he were to join society at large. I believe that the State has acted in this matter with good conscience and as if Mr. Newman were not a "human being", which he himself states that he is not.

Therefore, the State's position is that it has not violated any of the laws of this land and is not subject to the complaint brought by Dr. Paula Chapman.

Basically, if I may sum up, the State considers Mr. Newman to be property rather than a person and therefore not subject to legal codes of our land which govern the activities and rights of human beings. It is also my own opinion that the present hearing must consider all pertinent information relating to the "person" or "property" issue in order to properly rule on the due process issue before it.

The format for these proceedings will be as follows; the attorney for the plaintiff will put forth his arguments on each particular point; then, the State Defender will be allowed to respond. After all the arguments have been heard, each side will sum up their respective position in a short brief. The briefs and the court transcript records will then serve as the basis for my decision.

If there are no questions, I call on Mr. Darrow who represents the plaintiff to start these proceedings."

"Thank you your Honor. I hereby respectfully motion that Dr. Chapman's request, the basis of this hearing, for the release of the "person" known as Mr. Newman currently in custody should be granted immediately."

Chapter 13 Trial

"Your motion in this matter is denied" replied the Judge, "it has yet to be established that Mr. Newman is a person and therefore subject to human laws and therefore to such petitions that apply only to human beings."

"Very well, as you have clearly stated, your Honor," proceeded Mr. Darrow, "the crux of this case is the issue of whether Mr. Newman is to be considered "property" or a "person". In this regard, I will endeavor to present evidence and arguments which will prove that Mr. Newman is a person and therefore comes under our code of legal justice.

First, I wish to submit as evidence the journals and documents of one Dr. Joseph Newman. These items are presently in the evidence files of the Sheriff's Department and I'm sure that both the State Defender and your Honor are familiar with their contents. These will be the basis for part of my argument.

Wherein it is argued by Mr. Darrow that Mangor is genetically "human"

I call your attention to a clearly stated fact in the last journal of Dr. Joseph Newman that indicates that Mr. Mangor Newman was his son.

Based on this affidavit, Mangor Newman's genetic constitution is in fact at least fifty percent human or "*Homo sapiens*" in origin. Probably more so if we consider that his maternal parent also shared perhaps as many as 95% of genes in common with humans, an undisputed scientific

Chapter 13 Trial

fact, or the alleged cross fertilization could probably not have occurred.

By definition, human means: 'of, relating to, being, or characteristic of man.' In addition, 'Having or showing the nature, qualities, or attributes of a man'. Since Mangor by his birth certainly relates to and has genetic and other intellectual characteristics of "man", by these definitions alone he should be considered human.

I would also like to bring to your attention that while most people consider man as a single species, others have not and do not believe so today. Originally, even Linnaeus classified man and the anthropoid apes in the genus *Homo*. He included within this group *Homo sapiens,* or man, *Homo sylvestris,* or Orang, and *Homo troglodytes*, the chimpanzee.

Furthermore at present there are still two schools of human thought regarding the classification of humans; the monogenists who recognize only a single human species in the Hominid family, *Homo sapiens*; and the polygenists, who recognize three separate human species, Homo *indo-europaeus, Homo mongolicus and Homo niger.* Obviously, within the scientific community itself the genetic structure that comprises "human" is still a subject of wide debate.

Therefore, considering this degree of uncertainty among humans themselves regarding the precise classification of humans based on their genetic variation, I submit to you that Mangor does indeed have an appropriate

Chapter 13 Trial

genetic background to be considered a human being and thus should be subject to human laws and justice."

"Thank you, Mr. Darrow," the Judge responded.

"Now, Mr. Bryan, do you have a position with regard to the specific statements just presented by Mr. Darrow?"

Chapter 13 Trial

Rebuttal by Mr. Bryan: The case is made that Mangor is not genetically human

"Yes, your Honor, I do. The argument put forth by Mr. Darrow is quite appealing or so it seems on the surface. In the first place, Mr. Darrow has placed in evidence a book or testimony purported to be written by a Dr. Joseph Newman. Unfortunately, there is no proof that Dr. Newman is the author of the text and without witnesses to its alleged source; this exhibit cannot be considered a proper affidavit.

Our existing records containing an example of his signature on a document at a civil wedding performed in Reno is insufficient to prove the matter one way or another. Mr. Darrow also asks us to believe in the story as presented in that journal as the truth. Even if Dr. Newman were indeed the author, we would have no way to judge whether his statements are fact or fancy. What we do know is that we have in Mr. Mangor Newman a remarkable individual, but we have no way of knowing if he is in fact Dr. Newman's hybrid son.

I have also been informed that there are a number of other biological mechanisms that could account for the particular attributes of Mangor Newman, not the least of which is a normal, although unusual series of simultaneous mutations in his genesis.

However, for the moment and for the sake of argument I will accept the journal record of Dr. Newman. Even if what it states is the truth, Mangor cannot be considered

Chapter 13 Trial

human since he only "shares" some of the human genome, an element common to all existing life forms including bacteria, worms and jellyfish to name a few.

If we were to grant Mangor Newman "humanity" based on an incompletely shared human genome, where would we stop? What are the limits?

Mr. Darrow, as you yourself have so indicated the gorilla and indeed all of the anthropoid apes share portions of the human genome. Would we grant all apes in this country's zoos and research institutions a certificate of their humanity?

Where would the cut off be? We can follow certain common genes right back through snails, clams, worms and even bacteria. It would be ridiculous to have to formulate some way to determine "humanness" on the basis of the shared human gene pool. I therefore refute Mr. Darrow's present line of reasoning."

"Your Honor", asked Mr. Darrow, "In order to clarify this point, I hereby request that the State give us a definition of what is meant by the term "human". It would appear that such a definition would be extremely relevant to these proceedings."

"I'm sorry, but I have to deny your request" replied the Judge, "it is not the function of the State to define what is human and what is not. You will have to make your own reasoned arguments and these will be considered in forming our opinion."

Chapter 13 Trial

"I see," said Mr. Darrow dejectedly as he returned to his seat and began to sort through a number of documents on the table before him.

Wherein it is argued by Mr. Darrow that Mangor is a citizen by birth

Mr. Darrow then continued.

"Now, I would like to bring to your attention certain facts as alleged and presented in Dr. Chapmans's suit against the State of California. One of these is that Mangor Newman was born on United States territory and that his father was a citizen of the United States. Based on amendment 9, section 1 of the US Constitution which states:

All persons born or naturalized in the United States, and subject to the jurisdiction thereof, are citizens of the United States and of the state wherein they reside. No state shall make or enforce any law which shall abridge the privileges or immunities of citizens of the United States; nor shall any state deprive any person of life, liberty, or property, without due process of law; nor deny to any person within its jurisdiction the equal protection of the laws.

This, in other cases for which there is much precedent, he said waving the documents, makesMangor Newman a "citizen" of the United States of America by virtue of his

Chapter 13 Trial

birth alone. I also bring to your attention amendment 15, section 1 of the Constitution which states:

The right of citizens of the United States to vote shall not be denied or abridged by the United States or by any state on account of race, color, or previous condition of servitude.

Take note that this amendment specifically eliminates race or color, basically the genetics of its citizens, or whether they were somehow being held against their will and without their freedom in order to deny them of their rights as citizens.

Therefore, as a citizen, if not as a human being, I believe that Mangor must be given his rights and cannot be denied due process of law. I realize that Mr. Bryan has questioned Mr. Newman's parentage, specifically with regard to evidence that Joseph Newman, a human being, is his father. However, I am also prepared to present and will in fact present an overwhelming body of testimony and witnesses in science and medicine who will give us quantitative evidence that Mangor is indeed a hybrid, thus substantially supporting the documentation of Joseph Newman already in evidence.

Therefore, I move on the grounds that Mangor is already a citizen of the United States that your Honor must decide favorably on Dr. Paula Chapman's petition."

Chapter 13 Trial

"Mr. Bryan," said the Judge turning to him, "do you have any response to the claim that Mr. Mangor is already a citizen of the United States?"

Rebuttal by Mr. Bryan: Lack of evidence that Mangor is a citizen by birth

"Yes your Honor," Mr. Bryan said advancing to the Judges podium, "in my previous argument I stated my doubts about the validity of Dr. Newman's journal and its contents. Notwithstanding those doubts, I must now state that there is no legal record of Mangor's birth or of who his parents might have been. It might just as well be possible that another man, not a citizen of the USA might have been the donor of sperm used by Dr. Newman. There are also no living witnesses to either the birth or any who might be knowledgeable about Dr. Newman at the time of the alleged birth in the United States. Thus, there is no evidence of Mr. Darrow's claim and motion that Mr. Mangor be released based his alleged citizenship."

Mr. Darrow shook his head slowly and then continued.

An issue is raised by Mr. Darrow concerning how many human "genes" are required to be "human?

"Gentlemen, I would now like to explore another issue in this matter. Since your Honor has elected to refuse to provide us with an appropriate definition of what is to be considered "human" in this case, I would like to pursue its

Chapter 13 Trial

definition on behalf of my client. My esteemed colleague has also indicated that he has some question about the amount of genetic material that must be present in a person to be considered "human" and that in his opinion a human being must have one hundred percent human genes. Mr. Bryan, may I ask you a question?"

"Why, yes" replied Mr. Bryan.

"Do you believe in God?" queried Mr. Darrow.

"Why, yes I do", was the response.

"Do you believe in Jesus Christ?" continued Mr. Darrow.

"Well, yes I do indeed" replied Mr. Bryan.

"And Mr. Bryan, are you familiar with his parentage?"

"Yes" he replied.

"Am I not correct in stating that only Christ's mother was human?" asked Mr. Darrow, as the attention of the entire court turned toward Mr. Bryan.

"That is - correct" replied Mr. Bryan.

"Do you deny Christ's humanity?" asked Mr. Darrow.

"No, of course not!" responded Mr. Bryan.

"Then, I put it to you that Mr. Newman is just as human as that one other individual in which you and millions of other Americans believe so deeply. How is it that you can apply one standard to one person and another to Mr. Mangor Newman who is just as human?"

Mr. Darrow paused and looked about the court before continuing.

Chapter 13 Trial

"Your Honor," resumed Mr. Darrow, I hereby motion that the double standard being applied in this case be dropped and that Mr. Newman be rightfully considered "human" and released in accordance with Dr. Chapman's petition." The court was quiet as the Judge looked at Mr. Darrow, then Mr. Bryan and finally at Mangor.

"Your motion is taken under advisement by the court", responded the Judge.

"Your Honor," said Mr. Bryan, "I would like to protest this last motion on the grounds that Mr. Darrow has twisted the facts in this case to suit himself. He has tried to take our beliefs in God and apply them to---an animal."

"An animal is it then Mr. Bryan?" responded Mr. Darrow. "From what we know of Mr. Newman and the life he leads, I dare say that he is more humane, more sensitive, more considerate and much more rational than most human beings. Mr. Newman's intellect and reasoning power is also far superior to almost any other human being, including those present in this very courtroom! - and yet you refer to him as an animal?"

"The court will come to order," said Judge Dawson as he used his gavel for the first time, "let's keep to the specific issues at hand and not let this hearing degenerate into a personal conflict."

Both Bryan and Darrow glared at one another for a moment and then Mr. Darrow returned to his table and poured a drink of water.

Chapter 13 Trial

Rebuttal by Mr. Bryan: The issues concerning "genetics", "intelligence" and "humanness"

"Your right, your Honor" said Mr. Bryan. "We should not make this case personal. While it is my opinion that Mr. Newman is not a human being, I would now like to consider the matter of intellect raised by Mr. Darrow.

If indeed Mr. Newman is given "human" rights based on his intelligence, do we also not have to give these same rights to all creatures which are superior in this respect to some element of the human population?

How will we be able to slaughter cattle for our needs when they are more peaceful, rational and intelligent than the many humans that inhabit our mental hospitals and prisons?

Will horses and other animals be granted citizenship?

Will medical research come to an end because primates and other animals can no longer be used for research because they are smarter than some people?

I think that I have made my point and carrying this any further would lead only to more and more absurdities. I therefore propose that intellect, and specifically the intellect of Mangor Newman not be considered as an issue in this matter."

"Thank you Mr. Bryan" replied the Judge looking toward Darrow. "Have you any further arguments Mr. Darrow?"

Chapter 13 Trial

Mr. Darrow got up once again and faced the Judge.

"Yes, your honor, I would like to pursue one more point. Although I believe that I have proven by all reasonable measures that Mr. Newman is by his genetic constitution and intellect subject to human law, I would like to comment on an alternative position.

A question is raised by Mr. Darrow as to whether Mister Newman is a person or property?

Mr. Darrow approached the judge

"As I understand it, this hearing is being held to determine if Mr. Newman should be considered a person or property. As a person, we would hope that he would be immediately released from custody. While I believe and have stated that Mr. Newman is surely a person, let us consider for a moment that he might be classified as property. The question that then must be answered is just whose property he might be?

At this point Mr. Newman is in the custody of the State of California. However, by all reasonable standards related to Dr. Newman's document placed in evidence, it would appear that Mr. Newman, if considered property, is by right of his possession over a long period of time, the property of his three sons.

Since his sons are citizens and therefore cannot be denied their property without due process according

Chapter 13 Trial

amendment 9, section 1 of the US Constitution, Mangor Newman must be returned to them immediately!

Thus, regardless of the outcome of this hearing, whether considered a person or property, Mangor Newman must be released from state custody. Therefore, I now motion that Manger Newman be released immediately into the custody of Peter, Matthew and Andrew Newman".

The judge paused for a moment and then continued.

"Motion denied," he responded, "on the ground that there are peripheral issues in this case based on the same submitted evidence, namely possible murders and perhaps even kidnapping which must also be considered. As property, Mangor would certainly be responsible for any damage done to society in the past. However, as a person, he would also be responsible for additional charges based on the gravity of his actions which would then be the subject of another case.In either event, even if I granted the appropriateness of your motion, he could and would not be released from custody.

Mr. Bryan, before we take a break do you have any additional comments? "

"No your Honor", said Mr. Bryan

At that point the hearing was stopped for lunch and later, upon their return, the rest of the afternoon was taken up with a parade of witnesses both pro and con, dealing with facts and figures, which could and would substantiate

Chapter 13 Trial

one or another of the points made that day. By late afternoon, the Judge broke into the proceedings.

"We will now adjourn this hearing for the day and unless there is any problem, we will resume at ten o'clock tomorrow. Are there any questions?

Good! Court is now adjourned."

With that said, the courtroom was cleared and Mangor was returned to his quarters.

"Well," said Paula approaching Mr. Darrow, "it doesn't look too bad, does it."

"I think that it's too early to tell" responded Mr. Darrow with a twinkle in his eye,

"I believe that our arguments are sound, although I detect a slight partiality on the part of Judge Dawson. We'll have to see what tomorrow brings. At this point I'm only trying to give the judge as many reasons as possible to conclude that Mangor should eventually go free. Many of the reasons do not relate directly to your suit, but I hope that he will realize the inevitability of the outcome and that he will be inclined to grant your less complicated petition right now."

"How about Mangor," asked Paula, "will it do any good to put him on the stand as a witness?"

"I'm afraid not," replied Mr. Darrow thoughtfully.

"If this were a proper trial with a jury, we would probably gain from such a move. However, all of the parties involved know Mangor and have had extensive

Chapter 13 Trial

conversations with him. There's nothing more that I feel that he can add to this proceeding. I suggest that you go home and get some rest. I've got to get some things together tonight and prepare a brief for tomorrow. See you in the morning!"

Paula started out for home and some welcome rest. However, all the way to her apartment she could not help thinking about how curious it was that her God might be only half human - and that she had never even reflected on that very obvious fact. That night was a restless one for her.

THE SECOND DAY OF THE TRIAL

The hearings began promptly the next day with Mr. Darrow resuming his arguments right after the regular court business was out of the way.

"Gentlemen, this morning I would like to consider and enlarge upon our previous discussions with regard to whether Mr. Newman is now, or will ever be allowed to become a citizen of the United States of America.

Mr. Bryan has identified a seminal problem in that there appears to be no "legal" document or witnesses indicating that Mangor is indeed the son of Dr. Newman, a citizen of the USA. Thus, it is only Dr. Newman's journal which identifies Mangor as his hybrid son, and it is that testimonial which will have to be considered as evidence in this matter. I hope that this important document will be given the attention and recognition that it properly deserves.

The issue of whether Mangor has the right to become a naturalized citizen is addressed by Mr. Darrow

At this time, continued Mr. Darrow, I would like to pursue a somewhat different tack. As you are all aware the Constitution of the United States clearly indicates that it was written to cover the activities of "people" and I would and could not interpret its meaning in any other way.

Chapter 13 Trial

However, I bring to your attention part (4) of Section 8 of Article 1 of the Constitution of the United States which states that:

The Congress shall have the power...to establish a uniform rule of naturalization..

The Congress did in fact use this power and in 1952 passed the most recent law; the Immigration and Nationality Act. Now, I point out to all of you that this act provides for certain "aliens" further classified as "non-citizens" to be eligible for citizenship of the United States. That Act further removed all "racial" bars to naturalization in the United States.

Please take note of two very important facts: First, it states that an "alien" or in another word a "stranger", not necessarily or specifically referred to a "human being", can become a citizen. Secondly, racial characteristics or "genetic constitution" has been specifically eliminated as a requirement for naturalization.

It follows that even if Mangor is judged not to be a citizen by birth or by genetics, the Congress has established a route whereby he can become one by naturalization!

Since this is true, I hereby motion that your Honor act now and grant Dr. Chapman's petition to release Mangor Newman, clearly and obviously a resident of the USA who is eligible and entitled to become a naturalized US citizen, and whom furthermore as a resident of the USA based on

Chapter 13 Trial

Amendments 5 and 14 to the United States Constitution cannot be deprived of liberty "*without due process of law*".

Therefore, based on the United States Constitution, Mangor Newman, a resident of the USA cannot be held against his will unless and until he is arrested and given his due process.

At this time I propose that without his arrest and due process that Mangor, based solely on his residency is presently being held illegally in violation of the US Constitution and must be released immediately!

I must also point out that when the Constitution was written, everybody that was here was here at that time was only a "resident" and nobody was a "citizen" so that the fact of being a "resident" was the only requirement set forth by the founding fathers."

"Your motion is very interesting", replied the Judge looking at Darrow, "and will be taken into consideration. However, since at present Mr. Newman has not yet applied for a petition of naturalization or been granted such status, I cannot grant Dr. Chapman's petition on such grounds.

Mr. Bryan, do you have any comments with regard to the motion of Mr. Darrow?"

Mr. Bryan states that he does not believe that Mr. Newman's case is covered by either the Constitution or the Act of 1952

Mr. Bryan rose and approached the bench.

Chapter 13 Trial

"I'll have to take some time to consider this issue," he replied, "but in at least one instance there appears to be little merit in Mr. Darrow's position. That is due to the fact that in order to apply for naturalization the petitioner must have been lawfully admitted to live permanently in the United States. In this case, there is no such legal record or grounds to grant such legality."

"If I may add one thing, Your Honor" responded Mr. Darrow, "I would like to remind this group that Mr. Newman has been granted Ugandan citizenship, and being on United States territory and not wishing to return to Uganda where lawlessness reigns, Mr. Newman will immediately petition the United States of America for asylum. Thus, while there is a question of whether he is legally in this country, there is an enormous volume of precedent setting cases by which he surely will be granted asylum and permanent residency status in the United States. This request for asylum will then will be the basis for his petition for naturalization."

While the judge and Mr. Bryan individually considered Mr. Darrow's last point, most of the rest of that morning and afternoon were taken up with further witnesses that testified on the various points of constitutional law made in the arguments.

By the end of the day it was clear that the consensus of opinion among the court legal staff and lawyers present, with the notable exception of that of Mr. Bryan, favored the

Chapter 13 Trial

argument that the Naturalization Act of 1952, originally intended to eliminate discrimination between human races, did in fact by its specific wording, and as an unintended consequence, open a way in which a member of any other animal species could become a citizen of the United States of America.

The court was then adjourned.

That evening both Darrow and Bryan completed their briefs. The next morning, the lawyers were to submit then final briefs in the case and set out their arguments as precisely as possible.

Feelings in this case were running high throughout the world and the general mood did not favor the admission of a "non-human" to world society. The reasons were as varied as the individuals, their countries and sects involved, and were religious, philosophical and in many cases irrational or nonsensical in nature.

Paula and the sheriff had dinner together and discussed the events of the day. They talked late into the night and comforted one another.

Chapter 13 Trial

THE THIRD DAY OF THE TRIAL

The court was called to order at ten o'clock the next morning and the briefs were officially requested by the Judge who then asked if there were any additional summary statements prior to terminating the court aspect of the case.

"Your Honor," responded Mr. Darrow, "I would like to beg the courts indulgence and call Mr. Newman to the stand."

The Judge appeared to be annoyed and asked,

"Why have you waited so long before making this request?"

Mr. Darrow approached the bench slowly.

"Originally, I didn't feel that the issues involved would be either clarified by Mr. Newman's testimony, replied Darrow, "and I still do not. However, I think that it would only be right to have the individual most affected by the decision in this case have an opportunity to make a statement."

"Very well," replied the Judge, "Mr. Bryan, do you have any reservations or comments about hearing testimony from Mangor Newman as a witness for the defense?"

"Yes, I do have reservations, but no legal challenge that I can make and none that I think that your Honor would consider," answered Mr. Bryan.

Chapter 13 Trial

The testimony of Mangor Newman

"The court calls Mr. Mangor Newman to the stand," continued the Judge.

Mangor had been sitting facing the Judge and out of the direct line of sight of the majority of people in the courtroom. He slowly rose and stood for a moment looking at the Judge. The courtroom was absolutely quiet as this enormous being approached the stand, his dress making him appear even larger than he was. As a matter of court decorum, Mangor had been asked, and had agreed to wear a robe and sandals during the hearings. As he rose and approached the witness stand, he was quite a spectacle in his long white robe with loosely tied sash, sandals and, long black hair. All in all however, it was still hard for most in the courtroom to believe that this individual was not what he outwardly appeared to be, a massive and wild ape-like creature.

"Mr. Newman", the judge went on, "as you are aware, the State's position in this matter is that both your own and society's interests will be best served if you spend your remaining years under cloistered conditions and away from the mainstream of society. When this alternative was first presented to you by the Governor, you seemed to understand and concur. What we would like to explore at this time are the reasons behind your change of mind. Are

Chapter 13 Trial

you prepared to cooperate and testify for the benefit of this hearing procedure?"

"Yes", replied Mangor, and as he slowly looked about the courtroom his gaze fell upon Paula and they nodded to one another.

Mr. Bryan approached Mr. Newman now seated in an obviously too small chair next to the judge's podium.

"Mr. Newman", began Mr. Bryan, the key element involved in this hearing is whether or not your constitutional rights are being violated by your detention by authorities of the State of California.

The crux of the entire matter is whether you are entitled to such rights in the first place, and you have heard several lines of argument both pro and con on this point. What I would like to do now is explore your feelings in this matter. Mr. Newman, can you tell me why you want to become part of society at this late time in your life?"

Mangor replied somewhat defensively and his voice resounded throughout the courtroom.

"I have always been a part of society! Do you not see before you myself, my sons, and their wives?

Although I look different and have by choice lived an unusual lifestyle, I am just the same as you are, a product of and an integral part of human society. The question is not whether I want to join society, but whether that society should have the right to take away my freedom now that it sees how peculiar my nature is.

Chapter 13 Trial

If you deny my freedom because of differences in my nature, where will you go next?

Will it be the peculiarities of deformed babies or those with mental defects? Will it be skin coloration? Will it be the irreligious? or perhaps those without blue eyes?

If history serves me correctly, it is the Constitution which is our last defense, our protection against being discriminated against because of our genetic background or our infirmities or our special gifts for that matter."

"Mr. Newman," Mr. Bryan continued "remember one thing. By your own admission, you are not a human being!

Why should we then extend to you our constitutional rights? Is it just because you want them? If that is so, that's nonsense. There are many other intelligent animals who would want the same if they could ask for it. Are we to grant dogs such rights? How about porpoises or whales? How about the other primates who understand our language and can even communicate with us?

Mr. Newman, we are not insensitive, and in your case in particular we have tried to do what is best for all involved. We have merely made you a ward of the State. This is not unusual for we also apply similar criteria to human beings and in many cases put them in controlled environments as wards of the State for their own good."

Mangor shifted in his uncomfortable seat.

"Mr. Bryan", replied Mangor, I understand that some other "States" use these same criteria and apply them to

Chapter 13 Trial

those who have "different" political ideas or perhaps who have a sense of fairness not quite in vogue with those in power at the moment. Do you condone any position that the State might take with regard to "wards"? Cannot the State be wrong on occasion?"

Mr. Bryan looked away and then answered.

"Yes, I suppose so" replied Mr. Bryan.

"Mr. Bryan, I am not without my faculties, nor do I suffer from any disability which would cause me to need welfare or special attention. It therefore seems that in my case only the State's interest would be served by not letting me go free. And, although I have given the matter much thought, I am still not sure just what these "interests" might be.

There are millions of human beings out there in your society who need your care. They cannot cope with your society and are basely used by some elements of that same society which results in many human beings in such a state of misery that even I, an "animal" am ashamed. Mr. Bryan, why is it that you spend so much time with me and avoid the real problems of society? Is it because it's easy to delude yourself into thinking that you "care" by your actions here?"

At this, Mr. Bryan flushed and faced Mangor.

"Mr. Newman", replied Mr. Bryan, "please remember that it is you who are the focal point of this hearing and not me. In our Society we take care of many. Regardless of our

Chapter 13 Trial

feelings in any particular matter, and the slowness of our procedures, there will be no end if we do not make a beginning. Every case we try, every suit brought is a beginning and I do not feel ashamed of my role in bringing about justice, however small".

"Now", he continued, "unfortunately I feel that we have gone somewhat away from our original purpose here. The question put to you was why you wanted to become a part of human society after all of these years? and as I understand the position that you have clearly stated, you feel that you have always been a part of that society. Of course you realize that there are many others including myself that do not share this opinion, but this issue will be resolved as part of the larger question."

"Mr. Bryan" said Mangor, "I think that you have the wrong perspective about just where human beings fit into the scheme of things and perhaps you will bear with me while I give you a somewhat larger perspective to consider."

Mr. Newman presents a short history of the universe

"Mr. Bryan", began Mangor, "into the vastness of space and the infinity of time our Universe was born, accompanied by an incomprehensible violence some ten billion years ago. At that instant a cosmic cloud of matter formed and was sent hurtling in every direction by the great power which had been unleashed. Within the first five and

Chapter 13 Trial

a half billion years following that moment the stars formed out of the dust and the gases and in that way the galactic systems of the universe were created. There was light and there was darkness; there was cold and there was heat; and there were formed all of the elements required for life to begin. In the next one half billion years, bits of matter continued to coalesce around each of the stars, which together formed the many spiraling galaxies, and these in turn became the daughter planets of each of these stars.

In the infinity of space, in a single arm of one small spiraling galaxy, filled with myriad stars, one modest star called Sol started its evolution along with that of its family of planets which comprised our solar system. The third planet from Sol was earth which, about four billion years ago was new and still hot. In time however the earth cooled sufficiently and the water vapor which surrounded it began to fall to its surface as a liquid. The crust of the earth also hardened as it cooled and the rain which fell eroded and dissolved the rock mantle, and then formed rivulets, streams and mighty rivers as it flowed into the great depressions on the earth's surface. About three and a half billion years ago the earth looked much as it does today.

There were high mountains and great deep blue oceans and the sun rose each day on a blue sky filled with white clouds. The air was cool and the highest peaks were capped with snow. Each day the sun warmed the seas and vaporized the water which then rose aloft and was carried over the land where it cooled and fell to the earth once again. There were lakes and streams, ponds and rivers, canyons and plains. All were devoid of life. In the oceans, however, protected from the ultra violet radiation, a process was going on where chemicals associated with one another and recombined in an infinite number of ways in the warm saline waters.

In time a particular set of chemical compounds became associated in such a way that they could serve as the template for the formation of an identically complex set. Then they further evolved in such a way that the energy required for the strange and complex combination could be coupled to the energy derived from the oxidation or reduction of other chemical elements.

Finally, these primitive molecular life forms developed two important structures; one was a semi-permeable exterior membrane and the second was the ability to use the energy of the sun directly by the incorporation of special pigmented molecules which allowed chemical oxidation and reduction processes to occur within the confines of the membrane enclosed molecular complex. Thus, the bacteria and the first primitive plants evolved about two billion years ago. For the next billion years as oxygen filled the earth's atmosphere life continued to develop and radiate in every direction. Wind blown spores found their way into every moist place on the earth and began to grow and evolve under a wide variety of new conditions. Single

Chapter 13 Trial

celled animals developed in the seas and their colonies developed, in close association with the algae, into the multi-cellular colonial sponges and corals. On land the first primitive mosses and other plants developed from the simpler unicellular and colonial algae.

All of this occurred about one billion years ago. Molluscs evolved about 500 million years ago, fish and amphibians 400 million years ago, reptiles about 300 million and the first mammals about 200 million years ago. Monkeys developed fifty-five million years ago and the first primitive apes some forty million years ago. From the ape stem line, the gibbon developed twenty eight million years ago; the orangutan, eighteen; the chimpanzee, twelve; the gorilla, ten; and early man, some three and a half million years ago.

Of the primitive species of man, very few survived and one of the most recent species, *Homo sapiens,* has only been recognized on earth for some seventy thousand years. Farming and the development of domestic animals dates back no more than ten thousand years and corresponds to the development of the advanced human civilizations. Written language goes back only seven thousand years and the initiation of man's present religions only about five thousand.

As you can see, "humans" as presently defined have been on earth for only an insignificant time compared to their ape ancestors."

Chapter 13 Trial

"Your Honor", interrupted Mr. Bryan "we take exception to this history lesson on the creation of the world, and bring to your attention that there are millions of humans who do not believe as Mr. Newman does. We therefore implore the court not to take these statements into consideration in your deliberations."

"Mr. Bryan", continued Mangor, "please understand this. Whether or not you believe in my account of how the world evolved, regardless of how I came to be and what I am, I was in fact born.

I am here and I have every right to be here just as much right as you or any other living beast.

This is my world, my universe just as well as yours. And, if you think that I will give you my freedom willingly just because mankind has so decided you are wrong.

For the sake of my family and friends I have allowed others to seek redress through your legal system, but do not think for a moment that I will willingly remain your captive."

"Now, Mr. Newman" interrupted the Judge, "nothing will be accomplished by threats. We are here to try to understand the issues, so please control your temper".

"Mr. Bryan, you may go on", said the judge.

Mr. Bryan returned to his table and drank some water before slowly returning to the stand to resume his conversation with Mangor Newman.

Chapter 13 Trial

"Can you provide us" continued Mr. Bryan, "with any additional thoughts on why you should be granted citizenship and considered a legal "person"? We have all heard Mr. Darrow's arguments with regard to your genetic constitution and intellect and the possible legal arguments related to your parenthood."

"I think," replied Mangor, "that the basic question that needs to be answered is not whether I am to be considered "human" enough to become a citizen, but whether or not you must be "human" to become a citizen. Mine is the first such case of this nature, but I hope that you have understood my story of creation of the universe and that the several human species are only the latest steps in a long line of evolutionary changes resulting in "beings".

Just as your predecessors were not *Homo sapiens,* your descendents will probably also be so different that they will not be considered "genetically" human by today's standards. In addition, evolution may occur through catastrophic events rather than by slow and even processes and as a result, you may be faced with another individual such as me much sooner than you think.

There is also the possibility that some day soon, the human race may also be faced with perfectly rational beings, aliens or strangers of extraterrestrial origin. Will the democratic and free societies of the human races be too negative and exclude all other rational beings from the possibility of citizenship?

Chapter 13 Trial

What right does the newly evolved human race have to decide just how the ancient universe will be constituted?

My mother was just as much a "citizen" as you feel that you are. She was a citizen of the cold high mountains of the country now called Uganda by reason of her ancestors' evolution in that place and her heredity. Her family had lived there for eight million years before what you call human beings even existed. She was in her own right a proud, free independent and capable person. If not for the fact that her rights were violated and that she was kidnapped and brought to the United States as a captive and slave for display, I would not have been born and you would not be faced with your present dilemma.

You ask why I should seek my rights in this country. I ask you all to think about just whose rights were really violated? I am the product of violence, kidnapping, slavery and ultimately rape by the human race. How can you even question whether I am or am not a part of human society or whether I have any reason to seek my God given rights?"

"Gentlemen", continued Mangor as he stood up and surveyed the courtroom, "I'm not even quite sure just who should be on trial here today!"

Mr. Bryan backed away from the stand and the judge asked if he wanted to continue.

"I have no further questions at this time", responded Mr. Bryan," thank you for your cooperation Mr. Newman."

Chapter 13 Trial

"You may return to your seat Mr. Newman," added the Judge.

With that Mangor stepped down from the stand and with head erect, approached his table and once again sat down and was out of sight of most in the courtroom.

Mr. Bryan was the first to sum up.

Summation of Mr. Bryan

"Gentlemen, we have been involved in a forum in which we have presented and listened to arguments both pro and con and which will be used to evaluate a petition to release an individual now in the custody of the State of California. In my opinion, that petition which is based on the premise that the individual in custody is a "person" is moot if indeed there is no "person" involved.

The individual in question is an animal by his own admission; a highly developed animal with remarkable intellectual abilities. However, there has been no argument advanced thus far which supports a legal conclusion that this "animal" is a "person." We have no factual knowledge of Mr. Newman's parentage or genetic constitution, nor has there been any convincing argument that reasoning power alone makes an animal a "person" under our laws.

Mr. Darrow in defense of the petition to release this individual has made a pretty good case for the possible naturalization of Mr. Newman, but that remains the subject

Chapter 13 Trial

of a different hearing and is not relevant to the petition in question.

If indeed Mr. Newman is considered to be property, as I feel he is, the argument has also been presented that as property he belongs to his sons. However, there is no evidence for such a relationship. In fact, all legal records indicate that Dr. Joseph Newman, not Mr. Mangor Newman is the father of the children in question. I therefore believe that the judgment in this case is quite simple and that the State is and has not been depriving a person of his liberty without due process of law. Thus, I recommend that the petition of Dr. Chapman should be denied."

With that, Mr. Bryan returned to his seat.

"Thank you, Mr. Bryan," said the Judge, "the court will be pleased to hear from Mr. Darrow at this point if he has any further comments for the record."

Mr. Darrow rose and approached the bench

Summation of Mr. Darrow

"Thank you, your Honor," replied Mr. Darrow. "I believe that Dr. Chapman's petition is quite proper and correct in its allegation that a "person", namely Mr. Mangor Newman, has been and is presently being deprived of his liberty without due process of law. Since it is clear that we all agree that if Mr. Newman were considered to be a "person" that the petition for his release would be granted.

Chapter 13 Trial

Therefore, I would only like to review the evidence we have presented which deals with his classification.

As I see it, there are two prime issues which relate to whether or not Mr. Newman can be a "person" in the eyes of the law.

The first issue is whether or not Mr. Newman is to be considered a person by virtue of being "human."

The second issue is whether the only way to be a person is to be "human"?

I need not here reflect on the tragic circumstances and the unnatural act which resulted in Mr. Newman's birth and the consequences of that tragedy on the lives of the individuals involved. However, I do plead for the court to take all of this into account in deciding on how Mr. Newman will live out his remaining years.

Will it be in an institution or will it be in freedom in the society of rational creatures to which he has been denied through most of his life? We have presented evidence in the form of written testimony, witnesses and the results of laboratory analyses performed at the time of his treatment, which clearly indicate that Mr. Newman shares a large part of the genetic constitution of human beings and may in fact share no less a percentage of common genes present among various genetically different races of mankind.

Since there has never been a question of how many human genes are necessary to belong to the human race

Chapter 13 Trial

applied to any other citizen, it is patently unfair in this case to apply such an arbitrary and capricious standard.

We have also cited the fact that by virtue of the Naturalization Act of 1952 it is no longer required that a citizen must be a human being. That Act clearly states that an "alien" can become a citizen and expressly excludes his race or genetic background as a factor in determining the outcome of the proceedings.

Therefore, I implore the court to act favorably on Dr. Chapman's petition, since it is clear that Mr. Newman is entitled to be a "person" by both his humanity and by virtue of the implications of both the Constitution of the United States and the Naturalization Act of 1952."

Mr. Darrow then handed his brief to the clerk as Mr. Bryan had done and returned to his seat.

"Gentlemen," began the Judge, "I wish to thank you at this time for presenting your lucid arguments in this matter. Each of the issues raised will be given careful consideration and I now propose that we meet again in one month at which time the court will render a decision on the petition of Dr. Paula Chapman. Unless there is any further business, this Court now stands adjourned."

As the courtroom emptied and Mangor was being escorted to his rooms, Mr. Darrow turned and looked at Paula.

Chapter 13 Trial

"Well, Paula," Mr. Darrow began, "I think that we have a good case here and am somewhat more optimistic about it's outcome."

"I'm glad," replied Paula, but I can't help thinking about how Mr. Bryan's points were all related to matters of law. Legally, we don't know who wrote Dr. Newman's journals, nor do we have any evidence concerning Mangor's birth. The only legal wife Dr. Newman ever had was Mary White Newman and her sons are legally Dr. Newman's sons. You have to have a lot of faith in Dr. Newman's journal in order to believe a good part of our case. In addition, even if my petition is granted, Mangor cannot go free.

At that point he will probably be considered a "person" and will possibly be charged with a double murder and a kidnapping and who knows how long that will take to resolve. Incidentally, if Mangor is judged to be a person, Roger may then be charged with aggravated assault with a deadly weapon and with attempted murder."

"Anyway," replied Mr. Darrow, let's take things one step at a time. There are ways to solve each problem as it arises. Take care of yourself; I'll see you in a month."

"So long," said Paula to Mr. Darrow as Alan approached.

"How about lunch?" she asked.

"Sounds good to me," he replied, "it's exactly what I had in mind" he said smiling as they left the courtroom.

Chapter 13 Trial

THE LAST DAY OF THE TRIAL

The court was reconvened on the first day of March in the year 1985, Judge Henry Dawson presiding. He addressed the court:

"Ladies and Gentlemen, we are gathered here today to present to the public and to the participants in a suit brought by Dr. Paula Chapman against the State of California, the results of our decision in this matter.

The petition itself was very clear and stated that the State of California was in violation of the fifth and fourteenth amendments to the Constitution which state that no person shall be deprived of their liberty without due process of law. The State position was again clear in that, since no "person" was involved there was no violation of the law.

In order to properly evaluate this petition the court heard evidence with regard to the nature of the individual involved and his possible classification as a person.

The arguments on both sides were sound and it has proved to be a very difficult case to address.

We had to ponder many previously unasked questions;

What constitutes a human being?

What constitutes an animal?

Can an animal become a citizen of the United States?

Would a "throwback" or individual born to parents who were citizens, but who exhibited characteristics of pre-human elements, be a citizen by right of birth?

Chapter 13 Trial

We also had to ponder the questions of the rights and the responsibilities of both animals and "persons". I need not say that this entire area of law is without precedent and answers were not easy to formulate.

If each item were addressed from a strictly legal standpoint, our choices would have been simpler but not necessarily more just, and so any strict legal decisions were avoided.

There is yet another area of concern, only briefly touched upon during the hearings, which deals with the best interests of all involved, including the public and the country itself. The questions of maintaining a peaceful society and of not setting a precedent which would have grave implications for our hospitals, research institutions and animal farming and harvesting technologies are involved here.

Finally, we took into consideration the feelings of the individual, Mr. Mangor Newman, and what the outcome of the petition would have on him.

In summary then, the court feels that we have made a decision based not only on facts in this matter but also on a much broader interpretation of proper justice in his case.

The final decision

Therefore, it is the decision of the Supreme Court of the State of California that the petition of Dr. Paula Chapman be denied, and that the individual involved, Mr. Mangor

Chapter 13 Trial

Newman be confined to a state institution where he can live out the rest of his life in peace, humanely isolated from most of society, and with all of the conveniences present in order for him to carry on with a full and varied intellectual life as he has done in the past. In his confinement, he will be able to receive visitors and will also be able to communicate freely with society by means of his writings.

This has not been an easy case for a judgment to be rendered that is fair and just, and I pray that I have, in making this decision, served the best interests of society, of justice and of God.

This case is hereby concluded!"

Chapter 13 Trial

Chapter 14
Captivity

And so it came to pass that my father, an intellectually gifted human-gorilla hybrid, was declared to be an animal by law and so was to be held in benevolent captivity for the rest of his natural life, a period that is unknown since he is the only one of his kind.

The days following the trial were hectic and although Paula wanted to talk to Mangor immediately it was not possible for several months.

Finally, a visit with Mangor in his new home was arranged. Dr. Paula Chapman arrived in late afternoon and was brought to his residence by an armed guard.

"How are you?" said Paula enthusiastically.

Mangor, dressed in a robe and sandals, responded warmly giving her a gentle hug. He then called and arranged for them to have tea served in his study. While the tea was being prepared, Mangor showed Paula around his new home, a home befitting an individual considered to be a valued national treasure. The house adjoined the grounds of a large California state prison and had previously been the residence of the warden of the facility. It consisted of several large rooms, a bedroom, two guest rooms, a dining room, a kitchen, a workshop, an exercise room, a large shower room and most important a very large study with a

Chapter 14 Captivity

desk and bookshelves filled with books. On the walls of the study were several paintings that had been done by Mangor in his early years and had been kindly donated back to him by their present owners. In addition there was a large fireplace, several comfortable chairs and a table all made to fit the extra ordinary proportions of Mangor. Mangor, at nearly seven feet tall was not quite as tall as the tallest human, and approaching five hundred pounds, he was not as heavy as the heaviest human, yet with his large muscular body covered in long black hair he was indeed an imposing individual. Outside his spacious apartment there was a covered porch and a secluded treed outdoor area where he could sit or walk. The house had bars on the windows and the outdoor area was fenced.

Mangor's home was run by a staff that took care of his needs; its members were not unlike the staff at some zoos, but there were several differences. Many of them were trusted inmates of the prison who had chosen this service as part of their rehabilitation. There were also several civilian employees as well as three uniformed prison guards assigned to Mangor's residence. His foods of choice, within reason, were prepared and delivered to him, and the home was cleaned weekly the by civilian staff members. Unlike a prisoner however, Mangor was in complete command of his compound even though he was constantly reminded of his unfortunate position by the bars on the windows that looked out on the outside world. He could freely

Chapter 14 Captivity

communicate his needs and wants to the staff, and he could contact other individuals worldwide, via telephone, letters, and a novel asset, a computer. He also had complete access to all of the world's books and writings. All in all, his apartment was not unlike a luxury suite in a high quality residential hotel.

When the tea was brought in by a staff member, Mangor and Paula sat down at the dining table, and Mangor served her. They talked about the trial and its outcome and Mangor thanked Paula for all of her efforts on his behalf to gain his freedom. After some time, Paula asked if she could record their conversations for a writing a treatise on the trial and surrounding events she was writing. After Mangor nodded yes, she turned on her recorder and continued.

"Are you happy here?"

"Not really" replied Mangor, "but we will see what happens in time. As you see, I can have visitors including my family and in this single respect it is better than living in the mountains. However, it is at the expense of my freedom and that is something very difficult to deal with."

After some time, Paula abruptly asked, "Are you God?"

After what seemed a long period of silence, she then added, "I ask this only because many of the professed 'religious' people in the world believe that you are."

Mangor laughed quietly and then responded. "I am who I am and I am no different than any other living creature on earth. While I understand that there are many who think

Chapter 14 Captivity

that my unusual background is mystical, I assure you that I am just like you, a part of the same creation."

1
On the Nature of Creation

Mangor went on. "I have thought about the nature of our creation for a long time and have come to several conclusions. Would you like me explain my thinking on the subject?" he offered.

"Please do.", replied Paula, "since I as well as many other people are very interested in your philosophy, a philosophy that we unfortunately had little time to discuss during the rather hectic time of the trial."

Mangor continued, "Let me start by asking this: What child has not early in life considered the paradox of where he is in his own universe? As an entirely self-centered child looks about the world, the child sees that everything has a boundary and therefore cannot perceive any condition in which there is not a boundary. A child looks to the sky and sees the moon, the sun, and stars, and cannot but wonder about where they are in space. Then by virtue of a growing intellect the child considers the unimaginable. Everything must be somewhere and no matter how many boundaries are envisioned, and even if there were an infinite number of such boundaries the child logically concludes that even these must be contained somewhere-but where? Is it possible to be nowhere? It is at this point that all children

Chapter 14 Captivity

stop thinking about the problem in order to preserve their sanity."

Paula nodded in agreement, remembering when she had been one of those children.

Mangor continued. "As you know, current scientific thinking is that our observable physical universe was created about ten to fifteen billion years ago, initially in the form of energy at some specific point in space and that some of that energy has evolved into the matter that we observe all around us. Further, the belief is that this matter and energy have been expanding from this single point at an ever increasing rate. There is good scientific evidence for this. However, we are still left with the same simple paradox: Where are we? Even if we consider the possibility that there are multiple universes co-existing at the same time, the paradox still exists. What contains all of these universes? I myself, and just like many others, have no reasonable answer to this paradox, but nonetheless it is a paradox that lies at the center of all human scientific and metaphysical thinking."

Paula asked, "Do you believe that some other animals also think about this as well?"

"Maybe," responded Mangor, "but certainly in ways we do not presently understand, and may never be able to understand."

Mangor hesitated for a moment and then asked, "What then can I add to this discourse? Unfortunately, very little!

Chapter 14 Captivity

However, there is a growing consensus among scientists and philosophers regarding the nature of our creation. This consensus I can describe for you in some detail. Included are several basic elements related to our apparently unique and specific creation as are understood within the boundaries of certain human logical constraints. These include:

1.1

If a singular power created our universe, that power must be greater than its creation and therefore it cannot be contained within it.

1.2

If such a power exists outside of our universe, then it is also reasoned that our universe may not be unique and that there could be many similar creations such as ours.

1.3

No part of a creation can ever know the entire creation for that would require that the part be greater than the whole and therefore able to encompass the entire creation. Therefore, while mankind may gain insights into some of the laws of nature and speculate upon how the universe operates, by virtue of being a part of that universe, man will never be able to fully understand his creation.

Chapter 14 Captivity

1.4

Based on the natural law of entropy, creation can proceed only in one direction.

Since the paradox of where we are can never be resolved from information obtained about our known universe, children of all ages are correct in not pursuing this specific question further even though it always exists in the background and dominates everyday scientific and metaphysical thinking. However, these are worthwhile and rewarding inquiries. Scientific research can lead to an understanding and an appreciation of the mechanisms at play in our universe, and religion can make us feel better and calm the sophisticated inquiring mind that has evolved in mankind and some other animals."

Paula sat quietly for awhile and then asked, "Do you believe that the singular power that you mentioned residing outside of our creation could be a purposeful 'God'?"

Mangor contemplated Paula for a moment as she sat in one of his oversized chairs and then responded. "Surely I cannot answer that question with any authority since being a part of creation I also am confined within this universe and therefore I am unable to see any larger purpose even if one exists for this particular creation. As humanity and its science are both products of creation, man can never hope to prove or disprove the existence of a God that resides

Chapter 14 Captivity

outside of that creation. Nor can man ever hope to understand whether such a God is purposeful or not."

Mangor thought for a moment about how best to make this clear and then continued.

"In human terms, a mother can create a child with all necessary information for its growth and survival, but the mother can never be contained within the child. Conversely, a child can grow to understand some things about the mother, but he or she can never fully understand the entire nature of the mother. Is there a specific purpose for plants and animals to exist on earth? Not that I can see since on all of the other celestial bodies that we know about this has not happened and it seems to make no difference in their evolution. However, it is also clear to me that this particular creation is not random and chaotic in nature. There are natural laws that govern it, and by these laws, all of what we see today - including the presence of 'life' on earth - can be accounted for; there is no need to imagine any additional mystical, supernatural or purposeful interventions. In my thinking, the very existence of the universe itself is the miracle!"

Mangor was quiet for a moment and then considered thoughtfully. "What then can be said about the nature of our 'Creator'? First, that it is probably a single entity, since there is no evidence of any anomalies or contradictions in our present understanding of the universe that indicates more than one source of power. Second, that the Creator

Chapter 14 Captivity

has endowed the creation with a specific set of inviolable rules and regulations. Further, these rules and regulations are responsible for the creation's evolution, whether it is purposeful or not."

Mangor reflected for another moment and then added, "I suppose that some people could loosely interpret these assumptions as evidence of there being just one 'God' of the universe, one with a 'plan' for the eventual evolution of humanity. However, this also begs the question of the origin of that 'God' since by virtue of its very existence, in our minds it would also have to be a creation of yet another 'Creator'."

Mangor went on, "I also believe that all of the many supposedly purposeful 'gods' described by mankind throughout history are merely creations of the human mind and intellect. Given the enormous magnitude of creation, any gods who revealed themselves to man in antiquity and freely communicated with them, people living between a thin film of water and air on the surface of a speck of dust in the vastness of the universe, would have to be very small and insignificant gods indeed."

"Is there no presence of an almighty and purposeful 'God' on earth then?" Paula asked.

"No, not to my mind" responded Mangor.

"No 'prophets' of God either? No angels of God?" added Paula.

Chapter 14 Captivity

"No," replied Mangor, "and no saints and no heaven or hell either. These are just delusions of individuals who have strong beliefs and opinions as well as vivid imaginations. You may have wondered why the 'gods' look like men, behave like men and express many of the same emotions as men. This is entirely reasonable if you believe, as I and many others do, that these gods are cultural icons and products of the human mind and intellect with nothing to do with any divinity."

Then Paula asked, "Do you think that man is really no different than any other animal on earth?"

"Of that I have no doubt" replied Mangor, "since, as you can plainly see, I myself am a bridge between these worlds."

"These will be hard revelations to sell," she said. "Can you imagine a human society without a belief in a purposeful God on earth, without any divine restrictions, made up of people who did not fear eternal punishment for wrongdoing or look forward to eternal reward for proper conduct? A world where people did not believe they were special and superior to the animals?"

"Indeed I can," responded Mangor, "and it would not be very different from the one in which we live today. Humans, as well as most other social animals, would still live in extended clan groups as they always have, and in accord with certain universal tribal behaviors and restrictions that ensure the survival of both individuals and

Chapter 14 Captivity

the tribe itself. Very few social animals can succeed on their own outside of the confines of their own tribe. The specific tribal behaviors and restrictions often include:

1. There is only 'one' set of tribal rules. You shall understand and obey these rules or risk banishment.

2. You shall not attempt to change any of the rules of the tribe; if you do, this will be treated harshly.

3. You shall not deceive for when found out you will be punished by the tribe.

4. You shall observe all tribal traditions or be ostracized.

5. Honor your parents and elders or you will be reprimanded by them and also banished from the tribe.

6. If you murder you will be punished by the tribe.
7. You shall respect the moral codes of the tribe or you will be punished.

8. If you steal from a tribal member you will be punished.

Chapter 14 Captivity

9. You shall not harm a tribal member by spreading false rumors for you will risk being banished from the tribe.

10. If you exhibit greed and attempt to take your neighbor's goods, you will be punished by the tribe."

"Those sound very much like the ten commandments", said Paula.

"You are absolutely right!" Mangor replied. "Those commandments existed in some form long before mankind evolved, and these ten simple rules allowed primitive human tribal communities all over the world to develop less chaotically than they might have. These same rules, with some obvious modifications, are also followed by most groups of social animals today and allow them to live more or less peaceably with one another. With regard to mankind's specific belief that they are superior to all other animals", Mangor continued, "that belief is a good part of the reason for my trial and captivity and why I am here with you today. In the end, I guess it was easier for men to deny my humanity than to consider me an equal and in so doing admit to themselves that they were also animals!"

After a moment, Mangor continued. "I would also like to comment on one ancient 'Book', a book that is the root cause of my present confinement and the one that unequivocally states that mankind is unique among the creatures on earth and was made in the image of a specific

Chapter 14 Captivity

'God'. This is a statement and a belief that forms the current basis of much of human thinking about the universe and man's place within it. That specific document forms the bedrock of three of the largest religions presently adhered to by most of humanity and appears to be a historical novel but unfortunately nothing more. It is an interesting book that was clearly written by humans in antiquity as a tribal constitution within a patriarchal society and was based on long-existing myths and legends then prevalent in a wide geographic region on earth. It is also a book written at a time when, just like a child, a completely self-centered mankind considered itself to be superior to all other creatures on earth and its world was at the very center of the universe. In my mind, and in the minds of many others, this book was inspired solely by the human intellect and came into existence without any mystical or godly interventions. Clearly, the book had many authors and it was added to over time using a novelist's many communication skills and devices to develop a cohesive story. The book appears to have been compiled for the specific purpose of providing 'teaching' or 'instruction' on how to live in a well mannered and ordered society and thus avoid unnecessary conflicts. Moreover, when you think about the book's place in history, you have to consider that it was just one of many similar books written by men around the world at about this same time and for this same purpose."

Chapter 14 Captivity

"Are there, then, no God-inspired 'Bibles' in your thinking?" Paula asked.

"Not likely" responded Mangor, "as you now know where I stand on the subject of the presence of mystical talking gods on earth."

Paula made some notes and then asked Mangor if he was troubled by her questions. Mangor responded in a soft and kindly voice. "No not at all."

"You talk about the intellect a great deal. Can you describe the nature of the 'intellect' that you ascribe these human inventions to?" Paula asked.

"An interesting and important question", responded Mangor, "for it is the intellect of man that allows for both mankind's greatest achievements, and, its greatest disappointments."

2
The Intellect

"As I see it, the intellect can be described as the sum and substance of information stored in the form of memories, and its complexity clearly depends on the capacity of the brain that generates it. The brain itself is a complex information-processing organ with an innate logic, but it cannot see, smell, hear, taste or feel. For these things it relies on meaningful encoded electrophysiological signals that originate in a variety of out-of-brain neuronal sensors and without which input, normal cognitive abilities could not be sustained. Most scientists consider the brain to be the realm of the mind, the aggregate of all cellular processes that lead to an understanding of the nature of the surrounding world. While a brain operates only in the moment on a millisecond timescale, a mind is an emergent property of the sum of the activity of a brain over longer periods of time. During its lifetime, a brain continuously develops improvements in its circuitry and ability to store information as memories, eventually reaching a stage where it can store, retrieve and compare bits of information very rapidly. Thus, while a brain and its emergent mind are prerequisites for development of the intellect, it is only the level of acquired information stored in the form of memories at any given moment that constitutes the intellect itself. Metaphorically, you might consider the intellect as a garden of memories and the mind as the gardener who tends the brain's garden over time by clipping unwanted growth and nurturing other growth."

"I understand that memories can be stored and retrieved from specific places" said Paula, "but where in the brain does the mind reside?"

Mangor hesitated for a moment and then responded. "I believe that the mind is an ephemeral entity that exists throughout the brain's neural network and is a product of its moment to moment inter-neuronal signaling. As such, it is always active, always sampling all neuronal activities and modifications, including those associated with memories.

Chapter 14 Captivity

Remember that the brain is isolated in a warm, dark, and well-protected place and therefore relies completely on sensory inputs for information about its environment. These signals must be monitored continuously by sampling each sensor's input on a microsecond timescale in order for the organism to survive. Therefore, when you measure a person's brain bioelectric activity at any given moment, I believe that what you are actually measuring is the working of that individual's mind. Since a mind is a property of a functioning brain, it follows that a mind must also be present in all animals that have a nervous system and an ability to understand and store information about their surrounding world, a world in which they live, learn and exhibit behaviors appropriate for their levels of neuronal organization and suitable for their survival. In this regard there are a number of premises that I have found helpful in trying to understand the nature of the mind and the intellect. Perhaps they will be a help to you as well."

2.1

A brain is composed of cells, and its function is a product of the intrinsic signaling properties of each of the cells of which it is composed.

2.2

Chapter 14 Captivity

A brain can be very small and may consist of only two cells, a sensing cell that obtains information and communicates it to a second cell, which in turn acts on the message and produces an appropriate response. A brain may consist of specialized neurons or, as in the case of a sponge, even different kinds of cells.

2.3

A brain always operates in the moment and has the ability to rapidly store acquired information as memories, in the form of chemical-structural entities, which can later be retrieved. All memories are retrieved in the same moment without a specific timeframe to distinguish when they were acquired.

2.4

A mind is an emergent property of an active brain in which certain memories are selected and stored for long periods while others are discarded. A mind can not only search for, but also can create and store new memories. These memories may be false memories of a past that never happened, or memories of a future that may never occur.

2.5

Chapter 14 Captivity

Thinking is the temporal process by which a mind systematically searches for memories related to a specific task. A belief is the sum and substance of a systematic search of memories with a logical outcome. Since stored memories may be false, a belief may be logical but also incorrect. Thinking itself is an energy-dependent biochemical and biophysical process and is therefore subject to the universal laws of nature.

2.6
The intellect can be defined as the product of a mind at each stage of its development. While the quality of a mind is structurally limited by a brain, the intellect is quantitative and information driven. Thus, a good mind that has little information will have a relatively small intellect.

2.7
It is up to the mind using logic to distinguish between retrieved memories that are real and those that are false and to determine the proper chronological sequence and source of those stored memories. Failure to do so can result in a greatly troubled mind.

Paula looked quizzically at Mangor for a moment and then asked, "Are you implying that a sponge has a brain?"

Chapter 14 Captivity

"And a mind and an intellect as well!" responded Mangor. "There is really no reason to believe that a sponge can't think in some sense. A sponge is a loosely organized colony of cells that forms a complex animal. Each sponge species is different, but you just have to look at the varied and beautifully symmetrical forms that are built as a result of communication between their separate and specialized cell types to see the workings of sponges' minds and intellects."

"Getting back to the mind of man, many people believe that each of us has a soul, and some think that the mind may be the seat of that soul." said Paula, and she then quickly asked, "Do you believe in a soul?"

Mangor thought about the question for some time as he filled a pipe and then lit it. After a while he responded, "Yes, I do, but not in the way that many people do."

3
The soul

Mangor went on. "In many human cultures, there is a belief that there exists within each individual a unique soul, a metaphysical eminence conceived to be always present, to be separable upon death and to exist intact in the universe forever. Furthermore, they believe that such a departed soul may be animate as it was in life, and therefore able to communicate with the living. In some belief systems, it is posited that a soul is capable of being reincarnated. While there is no evidence that a metaphysical soul exists in the human brain, mind, or body, there is scientific evidence indicating the presence of a biophysical soul that corresponds in several ways to the metaphysical soul in these cultural belief systems. The scientific evidence of which I'm sure you are aware is based wholly on natural laws, on known aspects of brain function, and on the understanding that every change in the universe, however minute, reverberates throughout and perturbs the fabric of the entire universe for eternity. What has emerged from this analysis is that a well-defined biophysical soul certainly exists and can be measured; that its properties are unique to each individual; that it is terminated upon death and is immortal, in that its energy never leaves our universe. In addition, certain elements of a given biophysical soul can be reincarnated in the form of specific memories acquired by both mature and still developing brains and thus retained by others."

Mangor took another puff on his pipe and continued.

"It is also my belief that observation of certain aspects of the biophysical soul in the past served as the basis for various metaphysical soul myths prevalent today. The essence of the biophysical soul is that it is a biomagnetic entity. The living brain with its billions of neurons continuously produces bioelectric currents as a function of cell signaling. These bioelectric currents generate complex electromagnetic fields both within and outside of a brain.

Chapter 14 Captivity

The extra-cerebral magnetic fields can be measured and are the basis for my definition of a biophysical 'soul'. Such a soul can be envisioned as a biophysical entity that contains meaningful information in the form of a field of encoded and coherent magnetic emanations continuously being released by a functioning brain throughout its lifetime. Based on the first law of thermodynamics, which states that energy can be neither created nor destroyed, on the second law of thermodynamics which postulates the existence of entropy, and on the Einstein mass-energy equivalence equation ($E = mc^2$) concerning the relationship between energy and mass, it follows that the unique electromagnetic field energy continuously liberated by each individual into space can never be destroyed, only deconstructed as it is dispersed throughout the universe. By this biomagnetic definition, a biophysical soul can then be described and compared to a metaphysical soul. Thus, it follows that just like a metaphysical soul:

3.1
A biophysical soul is unique to each living individual, including humans as well as all animals.

3.2

Chapter 14 Captivity

A biophysical soul exists as a magnetic representation of the functioning of an individual's brain and mind and is continuously being released and dissipated within the universe.

3.3
The source of a biophysical soul, the bioelectric currents produced by communicating brain cells, terminates upon an individual's death.

3.4
Each unique biophysical soul is immortal in that the magnetic emanations from a living brain over its' lifetime continue to exist in some form in universal space for eternity.

However, let me add that unlike a metaphysical soul, it is also reasoned that while a biophysical soul is scientifically real and measurable, there is little likelihood that such an immortal soul liberated over a lifetime and continuously dissipated into universal space could ever be reconstituted as a discreet entity or be animated and able to communicate with a living human brain. Nor could it be implanted into the developing brain of another individual. These are simply illusions created by the human mind and intellect."

Chapter 14 Captivity

"I see by your reasoning that such a soul continuously being liberated over a lifetime may never be able to be reconstituted," Paula responded, "but is it possible that another human brain can recognize these emanating electromagnetic waves before they are too attenuated to make sense of them? I know that this has been scientifically tested many times and while most of the results are negative, on rare occasions there is a result that is statistically positive and as such forms the basis for still widely held beliefs in clairvoyance and the like."

Mangor mused, "It is certainly a possibility in the near time for we know that many animals and even some plant species have the ability to detect and make some sense of external magnetic fields. However, this ability even if it exists in humans could not account for myths concerning a soul as being a singular unchanging entity with the ability to be animated and communicate with living individuals over long periods of time. Just as observations of the physical world by early humans formed the basis of cultural metaphysical creation stories, I believe that observation of certain elements of the biophysical soul including its unique character while a part of each living individual; the stored memories of the qualities of that individual in others' brains and minds; and the apparent termination of that particular 'soul' upon death was the source of and formed the basis for human metaphysical soul beliefs."

Chapter 14 Captivity

Mangor then added. "In short, I believe that the concept of the metaphysical 'soul' was a natural construct of certain physical elements readily observable using normal human senses and then culturally processed and later mythically characterized by the human mind and intellect."

Paula and Mangor sat quietly for a long while and watched the flames in the fireplace as they moved and danced about.

It was getting dark outside and after some time Paula asked. "Are you tired?

"No" replied Mangor. "However if we are to continue, I suggest that we order some dinner and perhaps some spirits."

The dinner was brought in and eaten with relish followed by Mangor playing some unusual melodic pieces on his flute.

After a long time, Paula asked. "Can we talk about the origin of life on earth?"

Mangor replied. "Certainly"

Paula then asked: "What are your thoughts on the origin of life on earth and do you think that life exists elsewhere in the universe?"

4
Life on earth

"Again I must say that I don't have many more answers then you have already", replied Mangor, "but I have given

Chapter 14 Captivity

the subject much thought as you can guess. To me, the basic problem in understanding the true nature of life is that most people have arbitrarily defined just what 'life' is. The generally accepted definition requires that a living thing must meet all of several specific criteria in order to be considered 'alive'. Among these are included that it exhibit a defined cellular structure and have abilities to respond to stimuli; metabolize food; to grow, and to reproduce."

Mangor then continued. "Most life forms on earth as defined by these limited criteria consist of sophisticated bags of watery chemicals, a definition you may not think flattering but which also applies to both you and I. Amazingly, these bags of chemicals have in some cases reached such a degree of complexity that some of these entities have been able to understand many of the laws of the universe that have resulted in their very own creation."

"Think about that Paula!" said Mangor.

"I have" she replied, "but the question remains of just how did such complexity develop?"

"We'll get to that shortly", responded Mangor, "but first let's consider what these complex chemical entities are. If you look closely, you see that they are made up of the basic elements of the earth, and therefore it is with these substances and their intrinsic properties that the nature and origins of complex life forms must be sought."

"Again I see that there is no mysterious or supernatural force at work in your philosophy", Paula interjected.

Chapter 14 Captivity

"True, you have me there. I have no need for the presence of such a 'life force' in my thinking", said Mangor again lighting his pipe.

"The basic problem is how do we get from simple elements to the level of complexity generally ascribed to life forms, and more importantly, do we have to get there to describe 'life'? Many scientists have noted that individual 'chemical substances' have some of the abilities attributed to 'living things' such as the ability of crystals to grow and reproduce or to respond to chemical or electric stimuli or to store energy. The point here is that the 'life' that we describe only mirrors the cellular organization of ourselves and that such limited definitions ignore the much simpler ancestral life forms that must have preceded this degree of organization. Simply stated, if we are alive because we are a very complex chemical 'soup', then all chemicals on earth that make up that soup must also be 'alive' in some sense as even was the energy that was converted into their mass at a still earlier time. Based on this expanded definition, life on earth is not unique and therefore the entire universe is also alive with life forms of varying degrees of complexity. In answer to your question about the presence of living forms as currently and narrowly defined by mankind, I would be surprised if indeed similar complex life forms have not already developed many times and in many other places as it has on earth. Also, if this is the case, you should then also consider that all of the simpler life forms that I described

Chapter 14 Captivity

are still present on earth and continue to evolve even to this day. We don't see them as alive and undergoing 'Darwinian' evolution because of our arbitrary definition of what we believe to be a living entity. The following premises are the outcomes of my reasoning."

4.1
All energy and mass in the universe is alive.

4.2
The Laws of nature operate everywhere in creation and by themselves are responsible for the formation of both simple and sophisticated living things.

4.3
Sophisticated life forms like our own are probably present everywhere in our creation where there is a suitable environment and where they may be formed from any combination of "living' simpler chemical species at any time.

4.4
Life on earth and throughout the universe is continuously evolving from simpler to more complex forms. If all complex life forms on earth were to somehow disappear, they would reappear once again under the right environmental conditions,

Chapter 14 Captivity

Paula sat for a moment and then said, "I think that for the most part you have merely presented a semantic argument to explain that our current scientific definition of life is too narrow to suit you. By saying that all energy and substances are alive, you certainly argue against the need for a special 'creative spark' for life as we know it to be formed, but I'm afraid that it will not satisfy most people who strongly believe in such a 'creator'."

Mangor looked thoughtfully at Paula and then responded.

"Your right about my expanded definition of life, but isn't that already a basic tenet of many human religions; that all manifestations of creation including air, water and rocks are alive and that this creation was conceived and brought into being at the same time by an all powerful entity? I don't know what that entity was, but I'm actually in agreement with the concept that all parts of creation including energy and mass are all connected and equally 'alive'."

Perhaps we can change the subject a bit", said Paula, "and talk about the 'genetic code' that is considered to be a unique feature of living things on earth and is widely described as the 'code of life'."

5
The genetic code

Chapter 14 Captivity

Mangor walked to the fireplace, placed more wood on the fire and after poking around for a bit he continued, "The genetic code is certainly a wonderful construct made by complex living things. However, as far as I can see it is only a maintenance and repair manual. It was apparently formed long after sophisticated and complex cellular life forms had already come into existence. Think about it. If you are going to compile a dictionary, you must first have a language and its words and their meanings to be listed. Those words and their meanings must have formed before they could be encoded as a 'genetic code' and therefore represent earlier stages in the long evolution of sophisticated chemical 'life' forms, a period of time that may have been as long as one billion years. Of some interest, have you ever noticed that there was something important missing from the 'genetic code'?"

"What is that?" responded Paula.

Mangor went on. "If you brought a damaged car in for repair, its specific maintenance and repair manual would allow for repairing or replacing of each part. However, using the very same manual you could never hope to build the car in the first place. This is because there is no information on how to do it! The same is true for all of the genetic codes of 'life'. You cannot place a 'genetic code' for a specific creature into a chamber containing all necessary chemical materials and energy conditions, even if we knew what they were, and expect that the creature will be formed.

Chapter 14 Captivity

These codes were clearly written long after simpler non-cellular chemical life forms had evolved and therefore they do not have the specific information required for such a creation. With respect to complex cellular life forms, we can transfer a specific genetic code for one animal in the form of its isolated nucleus into another cell of a different animal that has been enucleated and in time obtain a creature representative of the new genome. It should be noted however that we must always start with an intact cell in these cases because of the clear and obvious fact that there is no information encoded within the 'genetic code' regarding details of the long evolutionary history of development of complex cellular chemical life forms and just how to build such a cell from its chemical constituents."

Paula thought for a moment and then said, "You are right now that I think about it, I can't see how I missed so obvious a fact? There is no genome that I know of has such encoded information about just how to form a living cell."

Mangor then continued, "Still the genetic code does give us some idea of how advanced carbon-based life forms evolved on earth. This can be deduced from the nature of code itself. In that code the only things that it' encodes are specific sequences of amino acids recorded and encoded as long chains of three "letter" chemical words. Today these sequences result in the controlled formation of specific amino acid chains that act variously as structural building

Chapter 14 Captivity

blocks, enzymes, hormones and transmitters. We also know from experimentation that amino acids must have been formed and were present in ancient oceans as a product of yet simpler substances by the action of lightning and other factors. In addition, the natural tendency of amino acids to form aggregations joined together by peptide bonds is also well known."

Mangor went on. "Here, allow me to speculate that our modern day carbon-based complex life forms now encoded in cells must have originated here on earth from these simple amino acids and their conjugates. In my mind, these amino acids were just as alive as any other chemical species and progressed to more complex forms based on aspects of Darwin's thesis of evolution and survival of the fittest entities. Although not generally considered, there's no reason that his theory cannot be applied to both the evolution of simple chemical 'living' species as well as to more advanced 'living' species. In a sense then, the genetic code while lacking specific information on how to form a cell, does give us a clue to the origin of cellular life in that it clearly describes these simpler amino acids and chains of amino acids as being our distant ancestors."

"An interesting proposition" said Paula smiling, "then based on your interpretation of the genetic code, our direct ancestors were amino acids, substances that we consume daily in chicken soup and other foods!"

Chapter 14 Captivity

"Precisely" responded Mangor." While I can add little more on this subject at the moment, based on logic and observation I suggest the following working propositions:"

5.1

The genetic code was formed after complex chemical life had already come into being on earth and as a result it does not contain information about how such complex life initially evolved or how to recreate it.

5.2

The genetic code provides us with some clues as to the chemical nature of our simpler ancestors and the environment in which they evolved.

5.3

Based on the genetic code, of all the possible chemical ancestors of life on earth, these appear to have been simple amino acids. While other chemical ancestors were possible, chance and conditions favored this evolutionary course.

5.4

Chapter 14 Captivity

The nature of the primitive environment in which these amino acids developed is also retained in the genetic code. That code recreates an environment of liquid water and 1% salt within a temperature range of between 1 to 100 C, and at a wide range of pressures as evidenced by the conditions under which present day life forms can exist.

Paula made some notes and then asked, "If we are creatures made from simple 'living' chemical substances purely by chance based on the operation of universal natural laws, doesn't it follow that our appearance on this earth at this time must have been predetermined by the act of creation itself? And if so, doesn't this bridge the gap between the act of creation and development of mankind by a God as the ancients believed?"

Mangor smiled and then said, "Yes, I have already alluded to this possible interpretation, but you must also understand that mans' presence on this planet is only one of many possible outcomes of natural universal laws. However, given an almost infinite amount of time for these forces to operate it is indeed likely that mankind in some form and in some place would eventually appear."

"This raises another issue" said Paula. "If sequences of actions through time are determined by natural laws governing the universe, it would seem to be impossible to change the course of such actions. On the other hand, it

Chapter 14 Captivity

seems clear to me that I can make choices and therefore that I have a 'free-will' and can decide to change the course of an action. Simply stated, if the future is predetermined by natural laws then free will cannot exist. Which is true?"

Mangor considered the question for a few moments and then responded.

6
Free will

"Many times I have thought about this question as have others through the ages."

After some hesitation, Mangor continued.

"It troubles me greatly. Does the past dictate the future and especially is it possible for an animal or a human being to choose to do something based solely on the will to do it? On the surface it appears to us that we do have choices to make and that our common sense then guides us in one direction or another. But is this true? Let me propose the following argument. If the universe could be stopped at any given instant and if we were all knowing and could note the position of every particle, its trajectory and associated energy field, could we not then predict where these particles and energies would be in the next instant? Here consider that an instant can be very short. Many agree that if the moment in time is short enough, such a prediction could be made."

Chapter 14 Captivity

Then Mangor asked, "Do you believe that this is possible?"

Paula thought for a moment and then replied, "Given your argument and if all of the laws of nature operate throughout the universe as you believe, I can accept the concept that conditions in a previous instant do in fact determine the conditions that will prevail in the next instant."

Mangor then continued, "If this is true, then by extension of this hypothesis back in time to the actions of human beings, each preceding moment in time must determine the next. Thus, in the instant that a human being appears to make a decision, that decision had already been determined by virtue of the previous instant. In other words, the choice was already made. Herein is another paradox! While a person feels that they have made a choice and can even visualize the outcome of that choice in the next moment, they made no such choice and therefore 'free will' must also be an illusion of the mind and its associated intellect."

Paula then asked, "If free will is an illusion, then how can we seem to be able to plan something in advance, to remember what we did and then recall what the outcome was?"

Mangor continued, "That is a function of memory. Memories are complex stored entities that do not exist at any given moment unless recalled using brain frequency

Chapter 14 Captivity

codes passed though specific fields of chemical-based physical structures associated with neurons. Thus, since the entire creation exists only in the present instant, all recalled memories also exist only in the present instant and there is no specific mechanism to separate them in time. As a result, the memory of a future decision to be made, the future action to be taken and the result of that action appear simultaneously and so the illusion of free will is created by the mind. As an example, many scientists use this technique to envision the outcome of an experiment in their minds. They go through all of the steps and all of the possible outcomes. In these mind experiments, the result of all possible outcomes are placed in memory and all can be activated anytime in the future. Since memories are always generated in the present, the mind and associated intellect are responsible for sorting out whether the memory of the future event preceded its outcome. This can create an illusion that a choice had been made in the past when in fact no such choice had been made - only a memory of that possible future choice. If the outcome is considered bad, then the memory of the wrong choice is thought to have been made. If the outcome is good, then the alternate memory of the right choice is considered to have been made. Let me summarize my thoughts regarding free will."

6.1

Chapter 14 Captivity

Every given moment in the universe is a function of the previous moment, and only the present moment exists. In this instant the past that brought about the present is gone and the future doesn't yet exist. Since each past instant determines the next, the present moment can never be changed. It also follows that absent a past or future, travel through time is impossible.

6.2

Time is an illusion and solely an arbitrary measure of change. If there were no moment to moment changes in the universe time itself would not exist. It has been calculated that at a very low temperature, all motion would stop, change would not occur and time could no longer be measured.

6.3

Free will is an illusion of the mind. Since one moment in time determines the next moment no change is possible.

6.4

Memories, no matter when they were formed, are always generated in the present. When memories

Chapter 14 Captivity

associated with a specific action as well as the outcome of that action are generated at the same instant without a time factor to distinguish them, no cause and effect can be differentiated. However, by searching stored memories, an illusion of free will is generated by the mind suggesting that one memory must have preceded another in time and that a choice had been made.

6.5

Memories of imagined future events can be stored just as well as memories of imagined or actual past events.

6.6

Memories do not exist at any given moment and must be recalled by passing the correct frequency code through a specific chemical-structural field of terminals between neurons. New memories in complex life forms are stored by altering the nature of these fields.

6.7

A specific memory is only elicited by the proper frequency code passing through the correct neuronal structural field containing that memory.

Chapter 14 Captivity

6.8

Without memories there can be no learning. The storage of multiple memories leading to good outcomes constitutes learning and as such serves to protect an individual without the need for free will.

Paula looked up and said, "I'm very tired. I have to go home. Are you telling me that my decision had already been made?"

"Yes" replied Mangor, "but your memory of a choice is only a memory of a possible future event which may or may not occur."

Paula looked questioningly at Mangor.

Mangor then said, "I know that you are tired, but let me try to give you a simple example. Imagine for a moment that sometime next week you will fly to London where you will stay at the Bedford Hotel. Have you ever actually stayed at this hotel before?"

"No" replied Paula.

Mangor excused himself for a moment, and upon returning several minutes later asked Paula, "Recalling your imagined trip, where will you go?

"I'll go to England", she replied.

"How will you get there?"

"I'll fly", she responded.

Then Mangor asked, "What hotel will you stay in?"

"The Bedford Hotel", she answered.

Chapter 14 Captivity

Mangor then said, "Moments ago you imagined a future trip to England, and at the present moment in time you were able to recall details of your planned trip. How could you do this? Where was this information stored?"

"I'm not sure" Paula said.

Mangor continued "Let me propose that you did it by creating in the past, a 'future memory' of a planned event that you could recall in the present. Your mind and its intellect then clarified that it was only an imagined trip that had produced that memory. You may never make this trip, but this plan was placed in your brain as a memory of a future event. Were you to make the trip, your mind would then suggest that by virtue of your free will you had indeed made a choice. On the other hand, if you did not make the trip, your mind would have then concluded that you again exerted free will and did not go. However, note that in each case the conclusion regarding the presence of free will can only be made *after* the event has occurred and does not *precede* it!"

Mangor shifted forward in his chair and gently massaged his brow. "Now that you mentioned it, I'm really very tired too. Look, I have an idea; at the State's expense there is a very nice room for you to stay here overnight. Why not get some sleep and then you can go home sometime tomorrow? I'll arrange for it right now. In the meantime I'll order something for us to eat."

Chapter 14 Captivity

Mangor then made the necessary arrangements and after a light supper they both sat looking at the fire for a little while before retiring.

"Good night, sleep well" said Mangor as Paula left the study. For a while longer Mangor sat quietly and thought about their conversations of this day and recalled so many similar conversations with many others through the years. Was this to be his future? Mangor thought. Locked away from the world forever? Mangor then got up and went to bed.

Next morning was bright and sunny and Mangor and Paula rose early and had breakfast on the patio. After breakfast they both sat for some time. Mangor smoked his pipe while Paula just closed her eyes and soaked in the warmth of the sun. After awhile, Paula opened her eyes and said, "I'll be leaving soon." "I hope that I'll be able to visit you again" she added.

"I'll look forward to it. Just let me know quite some time in advance since there are always people here running various tests on me. Not that I really mind, but it does interfere with my scheduling of other activities.

"Before I go", said Paula "we covered a lot of topics yesterday but I would like to ask for your thoughts on one more, the possible future of our creation?"

Mangor took another puff and put down his pipe. "Of course, you know that just like yourself and many others, my musings on the subject carry no more weight than those of anyone else. However, it's been fun to think about the possibilities. Also, I'm sure that you, like many of us may have at one time considered the possibility that there may not be any creation at all since the only evidence that you or I have that it exists is solely a product of our imagination. Nevertheless, I prefer to think that it does exist."

7

The future of our creation

Mangor then continued, "First let me say that there's no reason for me to doubt the outcomes of the many scientific inquiries regarding the laws of our universe that have been made. From them, we understand that the universe will probably continue to expand at a higher and higher rate of speed and into a space that we can't even imagine. As a result of the law of entropy, the future can only go in one direction and in the long term everything will eventually attain its lowest state of energy. At that time the temperature will have reached its absolute Kelvin level and all motion, all change and therefore time itself will cease to exist."

"A rather bleak picture from mankind's point of view", added Paula, "but fortunately it lies far in the future."

Mangor went on. "True. In a nearer timeframe and closer to home, we can expect our planet to be impacted by numerous other celestial bodies and if it survives these,

Chapter 14 Captivity

then conditions will change drastically as the sun slowly cools over the next 6 billion years. Life as we currently define it will end, but not life as I have defined it! Still closer in time and based on what we know about climatic changes and species on earth, mankind will end its brief appearance on earth and be replaced by another species. I think that we all make too much of the possible interaction of our species with the earth. It's probably true that mankind will eventually make his home on earth very different, but this is not a catastrophe. As I see it, it makes little difference whether the earth is covered in trees and advanced animal forms or with simple bacteria and fungi. It may make a difference to us, but we are here only for a moment in time. It is only in man's mind by virtue of his intellect that there was a purpose conceived for the formation of him, the earth and the heavens."

"Does God exist?" asked Paula.

"I believe that there is a Creator" said Mangor, and then continued. "Allow me to share some of my thoughts on the nature of this Creator and the possible future of our creation:

7.1
There is no discernable purpose for the creation of the "universe.

Chapter 14 Captivity

7.2

The creation can proceed in only one direction and governed by the natural laws of the universe it will eventually reach a point of its greatest entropy and time itself will cease to exist.

7.3

Mankind will cease to exist in the relatively near future and complex life forms may also cease to exist on earth at a somewhat later time. None of which appear to be of any importance to the creation.

7.4

The existence of our creation is both an enigma and a paradox. We believe in our minds that the creation exists, but cannot conceive of where it may be, why it came into being, or what its future might be? If the creation exists, logic also dictates that there must be a Creator that is greater than the universe that it created. Since no part of a creation can ever hope to fully understand the nature of a 'Creator' that exists outside of the creation, the question of the ultimate future of our creation can never be answered!

"Not a very good long term outlook" said Paula.

Chapter 14 Captivity

Mangor then responded, "If I were you, I wouldn't worry about it. Most of what I believe will happen is far in the future, and you should consider that I may also be very wrong in my conclusions."

"Comforting but untrue" said Paula, "I know that you are only voicing the opinions and beliefs of many scientists and other intellectuals. But more to the point, you describe a troubling universe without purpose where 'God' and 'free will' are illusions of the intellect and where there is no benevolent 'power' that looks after us in life and cares for our immortal soul after death. And, based on your arguments about the absence of free will, it is only an illusion that we think that we can make plans and then appreciate and feel better about the outcome after doing what we believe to be a righteous thing or to take pleasure in the actions of many other people who choose to band together to perform heroic or kindly acts."

Mangor took a deep breath and then said, "You asked for my philosophy and that's what I tried to give to you."

Then he continued. "But just like you, I'm not very happy with many of its obvious conclusions. I too prefer to live in a happier world that is filled with delusions and illusions, and in this unreal but comforting delusional state I also like to believe that I have a purpose on earth and that life on this little speck of a world matters in the universe. Also, I prefer to believe that I do have a free will and am able to decide what I would like to accomplish and where I

Chapter 14 Captivity

would like to spend the rest of my life. As you can see, in this respect I am again not so much different than you or the rest of humanity."

"Do we have the basis for a new religion?" asked Paula.

"I think so" said Mangor.

"We have a Creator that has performed what we can only describe as a miracle, the formation of our universe, and that Creator has also provided us with the ability to understand the natural laws that govern the creation. A Creator not angry or vengeful, nor benevolent or malevolent and that requires nothing from us in the way of sacrifice, worship or prayer. I call this new religion *'Naturalism'* and I believe that there are many followers of it already as many people of all faiths and even atheists are already practicing *'Naturalists'* who try to understand the natural order of the universe in which we live. Our Temples of learning are found within all of the nations and our Bible of natural understandings and its basic tenets are already written and can be found in libraries throughout the world. In addition, unlike other bibles that are based solely on events that occurred in a single cultural epoch, this bible is not static but very dynamic in that it is continuously being updated as each new piece of evidence of the nature of our creation is uncovered. *Naturalists* do not fear 'God' but instead live in awe of the Creator as they try to understand the immutable rules that govern our universe and its evolution."

Chapter 14 Captivity

"*Naturalism* it is then" said Paula.

Steps were heard as an armed guard approached Paula and Mangor.

"May I come again?" asked Paula.

"Of course" replied Mangor.

"May our enlightenment continue" she said, giving Mangor a hug, "I'll see you soon."

"Godspeed Paula, until we meet again."

Chapter 14 Captivity

Epilogue

After the hearing came to an end, my father was incarcerated. Although the physical presence of Mangor could not be denied, the story of his creation was determined to be unbelievable. He was therefore deemed to be neither a person, nor property, but only a valuable object to be locked away for safekeeping.

While the hearing ostensibly concerned itself with the fate of Mangor, on trial were also the faith and religious dogma of a majority of humanity that believed that they were unique in the universe, created by God in his own image, and given dominion over all other creatures on earth. The outcome of this hearing, later widely known as the second Scopes "Monkey trial" came as a great relief to many who sincerely held these beliefs, as well as those who used and harvested animals for a variety of purposes.

Shortly after the decision in the case of Chapman vs. California, Mangor was transferred to the custodial institution which was to be his home for some time. Within the next year the Society for Animal Rights and the Committee for Humane Legislation also sponsored Dr. Chapman's appeal to the Supreme Court of the United States, but the case foundered when the Supreme Court

Epilogue

refused to hear it on the basis that the State had complete jurisdiction in this matter.

The case for a petition for naturalization for Mangor was also dismissed on the grounds that he was an undocumented and illegal alien and thus not entitled to the benefits of the Naturalization Act of 1952.

Another case presented for his admission to the United States and eventual citizenship based on his seeking asylum was turned down because there was no legal evidence of where he was born and therefore no way to judge any basis for granting asylum.

Finally, in defeat, Paula returned to her anthropological studies and to the Chairwoman duties she had neglected for some time.

Mr. Darrow resumed his full legal practice and schedule and never championed a similar cause again.

Mr. Bryan unfortunately died of natural causes less than a year after the trial ended.

Roger Malek, who was responsible for many of the best and certainly of the worst years of Mangor's life stayed away from the hearings and took no further part in any proceedings. However, he was later charged and fined for hunting without a license.

The sheriff resumed his regular duties but his life was enriched by his unexpected and close relationship with Paula which had developed. This friendship was to continue throughout the rest of their lives as is common with people, even of vastly different backgrounds, who have shared an extremely emotional experience.

Judge Henry Dawson retired from the bench after the Chapman case and undertook a comprehensive study of the legal rights of animals. In 1987 he was instrumental in the sponsoring of a bill in the Congress of the United States to create an act establishing standards of care owed to certain types of animals with special consideration given to the Gorilla.

Not unnoticed by the many biblical and lay scholars who studied this unique case were the obvious parallels between the birth of Mangor and that of Jesus. Both were born in animal shelters, to a father named Joseph who was married to a woman named Mary. In both cases it was also noted that the child was only half human. Like Jesus, Mangor was a unique and a gifted visionary teacher and scholar who cared for his world and for the beings who lived in it. Both sacrificed themselves in the interests of humanity and were subsequently tried and convicted, without benefit of jury, of being socially unacceptable.

Now, while these curious circumstances were interesting, few really believed that there was any 'biblical' relationship between the two individuals. However, the religious and scientific scholars who pondered the question could not help but think how difficult it would be for a 'Messiah' or messenger who might one day come among us

Epilogue

in some unusual form in order to teach us the truth about our universe.

As for Mangor, after a period of some three years spent in captivity while waiting for the outcome of legal battles, and during which time he was subjected to tests of every kind, he once again left human society. It was on a bright sunny day in the spring of 1988, at the advanced age of fifty eight, his body crippled with arthritis and with only one lung and the sight of only one eye remaining, Mangor tore away the security bars on a window and headed toward the mountains to the east.

There were minimal efforts made to recapture him and those that knew him, as well as many who did not, were glad that he was free once more and hoped that when he finally died it would be in bravely fighting the adversity of nature in the high cold mountain wilderness which he loved rather than in the custody of mankind.

There were two rather strange facts about Mangor which surfaced during his detention and which have not been resolved to this day. Although there were records of Joseph Newman's birth and marriage, there has never been a record uncovered of his death or his ever having attended veterinary school or having received the degree of Doctor of Veterinary Medicine. In addition, although the records of the San Francisco Zoo were exhaustively searched, there was no indication that a female mountain gorilla had ever been purchased or that one was at the zoo in either 1929 or

Epilogue

1930. There was also no record of the appointment of a Dr. Joseph Newman DVM to the staff during that period. Perhaps these missing records will turn up in the future, but until they do, for some the origin of Mangor and his relationship to Joseph Newman will continue to remain a mystery.

However, there was never any doubt in the minds of my father Mangor, and our-selves. We believed the story as presented in Joseph Newman's journal was real for we had lived it. My brothers Mathew and Andrew also felt so strongly that the truth must be told, that they later both wrote extensive accounts of Mangor's capture, detention and trial as further testaments to the life lived by our father.

The new religion *'Naturalism',* the belief that there is a single Creator and that the universe is a self-existent, self-acting system whose characteristics and evolution can be explained in terms of natural science has continued to grow."

www.ingramcontent.com/pod-product-compliance
Lightning Source LLC
Chambersburg PA
CBHW071417170526
45165CB00001B/303